Rethinking Life at the Margins

Experimenting with new ways of looking at the contexts, subjects, processes and multiple political stances that make up life at the margins, this book provides a novel source for a critical rethinking of marginalisation. Drawing on post-colonialism and critical assemblage thinking, the rich ethnographic works presented in the book trace the assemblage of marginality in multiple case-studies encompassing the Global North and South. These works are united by the approach developed in the book, characterised by the refusal of *a priori* definitions and by a post-human and grounded take on the assemblage of life. The result is a nuanced attention to the potential expressed by everyday articulations and a commitment to produce a processual, vitalist and non-normative cultural politics of the margins. The reader will find in this book a unique challenge to accepted and authoritative thinking and new insights to research life at the margins.

Michele Lancione is Urban Studies Postdoctoral Research Fellow at the Department of Geography, University of Cambridge, UK.

Rethinking Life at the Margins

The assemblage of contexts, subjects and politics

Edited by Michele Lancione

Routledge
Taylor & Francis Group
LONDON AND NEW YORK

First published 2016
by Routledge

2 Park Square, Milton Park, Abingdon, Oxfordshire OX14 4RN
711 Third Avenue, New York, NY 10017

Routledge is an imprint of the Taylor & Francis Group, an informa business

First issued in paperback 2018

British Library Cataloguing in Publication Data
A catalogue record for this book is available from the British Library

Library of Congress Cataloging-in-Publication Data
Names: Lancione, Michele.
Title: Rethinking life at the margins : the assemblage of contexts, subjects and politics / edited by Michele Lancione.
Description: Farnham, Surrey, UK, England ; Burlington, VT : Ashgate, [2016] | Series: Cultural geographies | Includes bibliographical references and index.
Identifiers: LCCN 2015034757 | ISBN 9781472465757 (hardback : alk. paper) | ISBN 9781472465764 (ebook) | ISBN 9781472465771 (epub)
Subjects: LCSH: Marginality, Social—Case studies. | Sociology, Urban—Case studies.
Classification: LCC HM1136 .R48 2016 | DDC 307.76—dc23
LC record available at http://lccn.loc.gov/2015034757

ISBN: 978-1-4724-6575-7 (hbk)
ISBN: 978-1-138-54691-2 (pbk)

Typeset in Times New Roman
by Apex CoVantage, LLC

Contents

Figures

Contributors

Francisco Calafate-Faria is a sociologist and geographer whose research interests span urban political ecology, urban materialities, and urban transformations.

Cheryl Gilge received her PhD from the College of Built Environments at the University of Washington. Her research draws upon the philosophy of Deleuze and Guattari and examines mediated urban experience, the role of technology, and the tensions that emerge between social structures and individual practices that give shape to the social field.

Francesca Governa is Professor of Economic and Political Geography and teaches Urban Studies at the Polytechnic of Turin.

Eszter Krasznai Kovács is a postdoctoral researcher based at the Department of Geography at the University of Cambridge, working to capture and repoliticize the grounded effects of regulatory efforts and interventions made for environmental management, in particular for conservation. She has ongoing research examining the informal economies and 'codes' of farming underpinning access to land and natural resources.

Michele Lancione is an urban geographer and ethnographer interested in issues of marginality and diversity, activism, and continental philosophy. He is currently based at the University of Cambridge as Research Fellow of the Urban Studies Foundation. Michele is also Honorary Associate of the University of Technology, Sydney (UTS) and Book Reviews Editor of the journal *City*.

Jean-Baptiste Lanne is a PhD student in Human Geography at the University of Bordeaux Montaigne (France). He is attached to the Lab LAM ('Les Afriques dans le Monde', UMR 5115 CNRS). After a first work on festivity practices, his main academic interests now focus on everyday security interactions and individual tactics in the city of Nairobi, Kenya.

Gaja Maestri is a PhD candidate in Human Geography at Durham University (UK). Her current research focuses on residential segregation, public policies, and political mobilization in Italy and France, mainly with regard to the Roma and Travellers. She holds an MSc in Political Sociology at the LSE and a BA and MA in Sociology at the University of Trento (Italy).

Tawhanga Mary-Legs Nopera's creative practice affirms indigenous ways of being. He integrates his knowledge of *raranga*, or the traditions of Māori weaving, with live, image-based, and virtual performances. In particular, Tawhanga explores the spaces between mediatized, gendered, and sexualized identities. Through Māori creative research he asserts a means to expose, transgress, and transform bound terrains.

Darren J. Patrick produces creative and intellectual work at the intersections of queer, urban, and environmental studies. His research interests include queer ecologies, autonomous queer and feminist politics, institutionalized knowledge, and critiques of the globalization of radical theory. In 2015, Darren was a visiting scholar at the University of Bologna's Dipartimento Scienza Politiche e Sociali. He lives between Bologna and Toronto.

Matteo Puttilli is Senior Lecturer in Geography and member of the Laboratory for Social Geography at the Department of History, Archaeology, Geography, Fine and Performing Arts of the University of Florence. His research interests concern urban marginality and socio-spatial justice, environmental and energy sustainability, policies and practices of local development, geography education and didactics.

Kavita Ramakrishnan is Lecturer in Geography and International Development at the University of East Anglia. Her research focuses on notions of community, governance, and citizenship on the margins of the state.

Elisabetta Rosa is an architect and received a PhD in Territorial Planning and Local Development from the Polytechnic of Turin. Her research interests are focused on urban marginality and informal housing in Southern European cities and their relationship with urban planning and regulation. She is currently a postdoctoral fellow at the LabexMed, University of Aix-Marseille.

AbdouMaliq Simone is an urbanist and research professor at the Max Planck Institute for the Study of Religious and Ethnic Diversity; a visiting professor of sociology at Goldsmiths College, University of London; a visiting professor at the African Centre for Cities, University of Cape Town; a research associate with the Rujak Centre for Urban Studies in Jakarta; and a research fellow at the University of Tarumanagara.

Tatiana Thieme is an urban ethnographer and lecturer in Human Geography at University College London, having received her PhD from the University of Cambridge. Her research focuses on alternative cultural and economic geographies, particularly the politics of urban poverty and youth geographies of work in informal, underground, and 'hustle' economies. Her work also engages with environmental justice, especially urban political ecologies of waste.

Mark Tirpak is an urban planner, development specialist, and pracademic with a strong interest in *rasquachismo* – or the sensibility of making the most with the materials at hand, as informed by lean times and acquired aversions to waste. He is currently a PhD candidate with the University of Technology, Sydney (UTS) and a Visiting Researcher with The University of Texas at San Antonio (UTSA), Center for Cultural Sustainability.

Acknowledgements

As an Editor,
I would like to acknowledge the Authors
the Friends
and Family,

And then the benches, sidewalks,
beers, laptops –

Because without them
there wouldn't be a thing
to edit, not a single
word, line, or passage of
whispers, during an overheated
session in the rehearsal of
Tampa and its bay,
Carpet-stage
& AAG play.

I would like to bend this
page now, while I am writing,
bend the page you are holding,
and then give it to you,
cheap folio and nonsense,
but only when you look away
while you don't care,

So that someday you will open this and say
– *Look*, there was something
there. It was crafted within
the edges of borders –
Right at the margins.

***

To AbdouMaliq, Cheryl, Darren, Elisabetta, Eszter, Francesca, Francisco, Gaja, Jean-Baptiste, Kavita, Mark, Matteo, Tatiana, and Tawhanga for making this possible. Thanks to the Urban Studies Foundation for their support and to Katy Crossan from Ashgate, Faye Leerink from Routledge and Christopher Mathews from Apex CoVantage for their help. To Ash Amin, Francesca Governa, Marco Santangelo, Colin McFarlane, AbdouMaliq Simone, Edgar Pieterse, Stewart Clegg and Chris Philo for their mentorship and friendship. To my family. To Leo and life, for the poetry.

Introduction

1 The assemblage of life at the margins

Michele Lancione

Analytics of the margins

Being at the margin means to be situated on the other side of a border, while someone else is on the 'inside' somewhere more towards the 'centre'. Borders render the margins at the same time possible and visible, tangible and effective, embodied and felt. They are heterogeneous in their nature and forms but they bear the same message: one of expulsion enacted by multiple 'centres' to preserve their own authority and standing (Mezzadra and Neilson, 2013; Sassen, 2014). To say that the practice of bordering – of delineating what is right and what is wrong, what is in and what is out – produces margins, means essentially to state that through bordering specific spaces and people are produced as well: whether of the Mediterranean frontier and the migrants with their boats; a *maquiladora* and the exploited workers on their factory line; a xenophobic politics on gender and the stigmatized body of the queer; definitions of what is to be mentally healthy, a clinic and its 'mad', and so on. Not that these people and spaces did not exist before the powerful *oeuvre* of border making. But it is only through that *oeuvre*, which usually comprises more than one border and more than one margin, that their representation as marginalized with the attendant struggles and unbalanced power relations become visible, possible, and take place.

It is no surprise then that from Park and Stonequist's initial account of the 'marginal man' – where the 'border' was understood as the outcome of a conflict taking place in the mind of the migrant (Park, 1928; Stonequist, 1935) – social scientists have spent a great deal of effort in trying to define how particular borders work and on the effects that endure. The 'economy of knowledge' produced on the matter is overwhelming (Foucault, 2000, p. 216). Almost every study that is related to a group, a practice, a cultural trait, a gender, a socio-economic role, a space, and so on – that is not considered to be 'central', 'acceptable', or 'right' – could be defined as a work on marginality. The problem, as it has been argued (Del Pilar and Udasco, 2004), may well be that 'marginality' (as a concept) lacks scientific validity. It is simply too vague to be used as a category and thus not useful for sociological inquiry. Possibly because of its vagueness, most of the sociological and geographical literature tends towards 'adjective' marginalization: either speaking of socio-cultural, spatial, or economic forms of marginality (Billson, 1996). The rationale at play in this process is clear. Since 'marginality' is too broad, the concept is made manageable through unpacking specific (usually measurable) aspects that allow one to grasp, to define, and even to manage deviance from what is considered to be the cultural/societal/economic or spatial 'norm' (Bradatan and Caiutu, 2012).

With *Rethinking Life at the Margins* we move away from such compartmentalizations of knowledge – and of knowledge production – since we believe that it reduces the scope and strength of a critical scholarship (and politics) of the margins. In focusing *a priori* on a definition of what the margins and borders are, or in trying to contain their heterogeneity into strict theoretical boxes, a whole set of fundamental things – like the nuanced way power and affects work in the everyday life of people and their spaces – get dismissed or not adequately acknowledged. Moreover, as post-colonial scholars have shown, to define (from the centre) what margins and borders are without a committed engagement with the people and spaces of the margins, unavoidably ends up in the (re)production of stereotypical and disempowering knowledge (Spivak, 1990; hooks, 1994; Rose, 1997). In this volume we present an approach aimed at expanding this and cognate trains of thought, in order to acknowledge the complexities making up marginal contexts and subjects. A crucial starting point for us is to 'rethink' how we understand and come at terms with 'life' and its unfolding. The hypothesis is that for too long social scientists have not adequately dealt with the heterogeneity of life and, in doing so, they have missed out on potentially relevant aspects of how marginality comes together and evolves. Starting from this premise, the aim of the book is to offer a theoretical and empirical platform that stands as an invitation: to embark in a performative *minor thinking* (see what follows) built on the refusal of meta-narratives and *a priori* definitions on marginality; on a post-human, grounded take on the assemblage of life; on attention to the potential expressed by everyday life's articulations; and on a commitment to produce an alternative cultural politics of the margins. This book is provided as a source for inspiration to implement a critical rethinking of life at the margins; to be dismantled and used; to be challenged and to push forward.

In order to clarify these last points, and at risk of being too simplistic, we can identify three analytics at play in how social scientists have been dealing with the margins so far. These are neither schools of thought nor methods. The structural, grounded/relational, and post-colonial analytics signal tendencies and blueprint for research. Briefly recalling them allows an appreciation of the particular take on the margins proposed in the present volume.

The first group concerns what we can call the *structural* analytic. Through it, the margins are conceptualized as an outcome of structural processes of dispossession, exclusion, and exploitation. One recent example in this sense is Loïc Wacquant's successful account of 'advanced marginality' (Wacquant, 1996, 1999, 2008). Wacquant offers a macro-level analysis of urban marginalization through the lens of two concepts: 'race' and 'class'. He does an excellent job of identifying the structural causes that have brought about an urban marginalization that is moving away from its Fordist forms, clearly showing the different patterns followed by cities in the EU and the US. However, 'race' and particularly 'class' are used by Wacquant not only to identify the patterns of advanced marginality but also to explain them and to offer solutions for them (Tissot, 2007). This is what usually characterizes the study of marginality from a structural perspective: one variable, considered to be structurally more important than others, is used to identify, explain, and tackle marginality-related issues. This kind of approach is most evident in studies of urban marginality based on traditional political economy, where the 'hidden' force of capital is both cause and explanation of marginalization (Castells, 1977; Harvey, 1973). It maintains its validity even in more recent and less orthodox accounts focused on the neo-liberalization of city space and politics (Mele, 2013); on social exclusion dictated by the interplay of the state and the market (Mingione, 2004); through global macro-economic forces that shape the restructuring of urban peripheries in the global South (Davis, 2006); and through neoliberal policies of containment towards the poor (Mitchell, 1997; Smith, 1998), to cite but a few examples. Reading the margins through a structural analytic can offer valuable macro-insights into the broader dynamics of marginalization. However, it is arguably limited by its frame: it cannot see borders other than the one constituted by the structural variable that is put into focus.

The second analytic is what we may call *relational/grounded*. With these terms I intend to signal approaches that prioritize the relational level of day-to-day interactions and makeshift reactions and processes in their depiction and analyses of the margins. This grounding can be more or less blind to 'external' or 'structural' dynamics. There are writings that one could name, following Ash Amin (2013), 'human potential' urbanism, focused on the creativeness and resilience of the people living at the margins (Neuwirth, 2007; Saunders, 2012) that seem to lose touch with the macro processes influencing marginal contexts (for a critique, see Obeng-Odoom, 2013). A more balanced view is offered by literature combining a rich ethnographic account of marginality alongside in-depth readings of the economic, cultural, and social forces related to the condition of specific spaces and subjects. In Geography, Ayona Datta's *The Illegal City* is a recent example of this kind of scholarship. If Datta gives a very lucid account of the legal, judicial, and cultural contexts of Delhi, 'within which the everyday negotiations of law and illegality acquire meaning and significance in squatter settlements' (Datta, 2012, p. 14), then at the same time she explores those settlements from within, offering a rich ethnographic account of Delhi's marginal spaces, which are rendered clearer through insight into the wider context. The 'poor' here are neither framed *a priori* nor celebrated *per se* but rather situated and understood from the place-specific dynamics and logics in which they live. The literature offers similar accounts across a variety of fields: feminist studies (think of Radcliffe's work around Indigenous women in Ecuador, who situate their claims for land in a broader field made up of traditions, law, and prejudice; Radcliffe, 2014); research on the city of migrants and related issues of hospitality (Darling, 2011); studies of populations affected by mental health issues that cast new light on related spaces of care (Parr, 2000); research around drug and alcohol abuse at the margin, enabling an understanding of this issue in non-stigmatizing terms (Bourgois and Schonberg, 2009); homelessness scholarship that confronts homeless people's street life from a performative perspective (Desjarlais, 1997; Duneier, 1999; Johnsen, Cloke, and

May, 2005; Gowan, 2010); investigation of 'slums' encompassing and challenging canonical meta-narratives, both in cities of the North (Venkatesh, 2008) and the South (Weinstein, 2014); and more. Although some of these works are focused on the micro, at risk of forgetting 'macro' dynamics and processes of oppression (which, for instance, is Wacquant's criticism of Duneier's work; Wacquant, 2002), the best of this scholarship is able to portray the margins from the inside out and back again, connecting local issues and potentials with broader socio-political dynamics that are not there to frame the narrative but to enhance a street-level understanding of marginalization.

Last but not least, the literature on marginality has a huge debt to the *post-colonial* analytic. Although post-colonial studies can take either a structuralist or a grounded approach to the field, what defines and differentiates these from previous analytics is their explicit commitment to produce a non-normative kind of knowledge and to provide a critical and self-reflective account of the 'other' (Barnett, 2006). Central to the post-colonial analytic is a concern with cultural discursive formations – namely both with the practices of producing meaning and the production of knowledge through language (Hall, 1992). In this sense, for the post-colonial critique the experience of being 'marginal' is a 'consequence of the binaristic structure of various kinds of dominant discourses, such as patriarchy, imperialism and ethnocentrism, which imply that certain forms of experience are peripheral' (Ashcroft, Griffiths, and Tiffin, 2007, p. 135). Post-colonialism is deeply politicized because its aim is not only to describe these cultural practices of border making but also to challenge them and to provide alternative practices and knowledge (hooks, 1989; Ferguson et al., 1990). Perhaps the most evident and powerful example of such politicization is in the notion of the 'subaltern'. The Subaltern Studies Group introduced the Gramscian notion of the subaltern not only to account for, and to write differently for, the marginal subjects in Indian history, but to rewrite that history entirely in order to fight dominant Western narratives. The subaltern cannot speak – says Spivak (1988) – if s/he does not learn how to speak the language of the colonizer, but at the same time s/he can construct differential knowledge to challenge the dominant culture (for a concrete example in this sense, see Gunew's account of Aboriginal literature in the Australian context; Gunew, 1994). As Huggan puts it, what drives the post-colonial approach to the margin is 'an oppositional discursive strategy that flies in the face of hierarchical social structures and hegemonic cultural codes' (Huggan, 2001, p. 20). Post-colonial geographers have shared this ethos in approaching and investigating the margins (Sidaway, 2000; Nash, 2002) and the same can largely be said of feminist and queer geography (Sharp, 2009). Within the latter, particularly in situated knowledge approaches, considerable effort has been put to theorize the researcher's own positionality in the field and ways to make research at the margins ethically sound and politically relevant (Katz, 1994; Rose, 1997; Pain, 2003; Besio and Butz, 2004). What post-colonial approaches and feminist geographies have in common is that the social construction of knowledge around the margins is not taken for granted but questioned, in order to transform the unequal and oppressive power relations characterizing it (Staeheli and Lawson, 1995; Arabindoo, 2011).

These three analytics are far from exhaustive. Many approaches to the study of the margins have risen from disparate academic fashions, cultural paradigms, and contextual traditions (O'Connor, 2001). Moreover, many scholars have mixed one or more of the above analytics. In geography this is evident in the seminal works of Gibson-Graham (1994), Cindy Katz (2001), Doreen Massey (1993), and Ananya Roy (2009), to nominate just a few among many who have combined the critique of socio-economic structures, committed fieldwork, and post-colonial ethos to question borders and their margins. Nonetheless, outlining these analytics allow us to clarify the ground upon which we have decided to write this book. With *Rethinking Life at the Margins* we are interested in developing a critical approach based on the latest advancements in social theory, offering a more nuanced understanding of the dynamics of bordering and marginalization. This is not done in opposition to what has come before. Rather, the approach that we advance in the book is linked to a fourth analytic that is largely built upon the *relational/grounded* and *post-colonial* take on the margins. This is what one could call a 'vitalist' analytic because its aim is first and foremost to account for the many lives that make up the social and hence the many forms of life that construct, sustain, demolish, and rebuild the borders we so incessantly study and write about. We argue that reapproaching the margins through a vitalist analytic allows both a focusing on issues of marginalization in more nuanced and politically relevant ways *and* to see things previously unseen.

In what follows I will first introduce the fourth analytic and illustrate how it relates to – and how it aims to expand – the three just outlined. Second, I will introduce assemblage thinking, namely the approach that, at least in human geography, has been more effectively implemented to study the margins from a vitalist angle. I will then illustrate the two points that we consider at the centre of the *oeuvre* for rethinking marginality, namely reassessing the contexts and subjects of the margins. From this reassessment it will then be possible to highlight the ethics and cultural politics that fuel our common effort.

Towards a vitalist approach to the margin

In recent years social scientists have exerted considerable energies trying to conceptualize how the social world is made up by (what has been usually conceived as) non-social matters: non-human beings, artefacts, natural forces, atmospheres, and so on. These things have of course always been part of the social, but they have rarely entered social theory's concern. Although there have been exceptions (think of Tarde's monadology; Latour, 2001; or of Reclus' anarchism; Clarke and Martin, 2004), it is only more recently, and in particular with Haraway's cyborg feminism (1991) and Callon, Latour, and Law's Actor Network Theory (ANT) (Law and Callon, 1992; Latour, 2005), that the non-human takes centre stage. These cognate approaches in a variety of disciplines – think of 'socio-materiality' in Organization Studies (Orlikowski, 2007) or the more-than-human in Geography (Whatmore, 2002; Philo, 2005) – bring to the fore three main points. First, there is attention to non-human agencies and the role they play in changing the state of affairs of things. The social world is 'repopulated' with non-human matter, leading to the rewriting of established social categories and knowledges around subjectivity, space and place, social relations, and organizations. Second, things are investigated in their unfolding and not from a pre-established paradigm or *a priori* explanation. The cause-effect model is rejected in favour of a processual approach aimed more at showing how things are organized than at explaining *what* things are (Murdoch, 2006). Third, there is a general call to overcome the limits of canonical forms of representation, in their being exclusively human and rational centred, in order to account for the power of emotions (Bondi, 2005), shared affects (Anderson, 2009), and the unconscious (Thrift, 2005).

The stances expressed by ANT, socio-material, and non-representational approaches form the ground of the fourth analytic this book relies upon. This is what we can call a 'vitalist' or 'post-human' analytic because it not only acknowledges non-human agencies, but it also moves human and non-human *onto the same ontological plane* (this is what, in the end, distinguishes a 'vitalist' scholarship from the 'more-than-human' approach). Vitalism puts a stronger stress on the need to decentralize the human and its assumptions in order to reimagine *how* the social comes together and falls apart. Things – human and non-human – are approached through a rejection of subject-object distinction. What gives things prominence is not their being a 'subject' or 'object' of action but their being 'alive', namely the fact that they share a common ground for their existence.

Building on the Spinozian idea of *nature* and of *conatus*, Deleuze has called this ground '*a life*': a plane free from subject-object distinctions and permeated by a power that does not belong to something or that can be found in something in particular (Deleuze, 2001, p. 26). The power of being alive and becoming – namely the immanence of a *life* – is of everything, in everything, through and over everything. It is, as Bennett puts it, 'a restless activeness, a destructive-creative force-presence that does not coincide fully with any specific body. A *life* tears the fabric of the actual without ever coming fully "out" in a person, place, or thing' (Bennett, 2010, p.54). The immanence of a *life* is what human and non-human share. It is their common ground, which 'occurs before and alongside the formation of subjectivity, across human and non-human materialities and in-between distinctions between body and soul, materiality and incorporeality' (Anderson and Harrison, 2010, p.13).

Importantly, this common ground is at the same time about potential and actualisation. As Deleuze and Guattari would have it, things acquire their perceptible status but they also bear the potential of becoming something else (Massumi, 1992; DeLanda, 2005). For Deleuze and Guattari this potential is not something volatile or unintelligible but it is as much real as the actual form. The actual and the potential take place at the same time and in the same space, in the becoming of things through a *life*. In other words, everything is real and therefore everything is (potentially) possible: 'The real is not impossible; on the contrary, within the real everything is possible, everything becomes possible' (Deleuze and Guattari, 1977, p.27). A vitalist approach is therefore not only interested in understanding life as a shared feature of human and non-human, and in tracing and depicting its articulations, but also in encompassing the actual form of things to take into account their possible, differential, becoming.

The most profound point of rupture with the analytics listed above is precisely this *holistic* and *in-potential* understanding of life. On the one hand, as Braidotti has highlighted, a post-human account rejects 'the opposition nature-culture and stresses instead the self organizing (or auto-poietic) force of living matter' (2013, p. 3). On the other, vitalism is not only interested in 'this or that plane, nor this or that realized system of relations, but [in] the potential to produce planes, the "planomenon" and our capacity to think or encounter that potential' (Colebrook, 2004, p. 6). This does not mean discarding matters of structure, relations, class, gender, colonialism, and race, to cite just a few of the categories at the centre of the previous analytics. It does mean that one needs to reject their status as 'social fact', or interpretation as only human dependent. The task of a critical vitalist/post-human social science is to reimagine these and other matters taking into account the countless non-human elements and entanglements making up life in order to reimagine politics as well. Structures, relations, and cultural politics are then reconsidered from the immanent, the potential, and the ubiquitous nature of life.

Thinking through assemblage

In order to approach the world through a vitalist lens one needs a different way of thinking and performing research. What is needed is a *methodological sensibility*: if 'vitalism' signals a particular take on *a life*, then we need a set of propositions, insights, and theoretical tools to enable us to navigate the vitalist plane. The scholarship that in recent years has been produced under the umbrella of 'assemblage thinking' holds the most promise to this end. This is because 'assemblage thinking' is interested in tracing human and non-human affiliations in their becoming, highlighting both potentialities and limitations, without giving prominence to the 'human' component in the process. It is worth highlighting that in this book we do not consider assemblage thinking a theory but, as Law has also argued for ANT/material semiotics, we interpret it as an approach, 'a sensibility to the messy practices of relationality and materiality of the world' (Law, 2009, p. 142). This sensibility is what interests us in *Rethinking Life at the Margins*. We read the notion of assemblage (or *agencement*) as an invitation to a particular kind of processual thinking, one able to read the margins in their post-human entanglements without reducing their predicaments to pre-established theories or narratives. The people and spaces of the margins are confronted from the thick, grounded description of humans and non-humans in their making (McFarlane, 2011a), with the double political aim of overcoming the formalism and structuralism of canonical thinking (Jacobs, 2012), and of acknowledging the actual and potential multiplicity of any composition (Dewsbury, 2000).

Greenhough has pointed out that we can distinguish three ways by which assemblages are implied in the current literature: the anticipatory, which sees assemblages as a way to depict the world yet to come (Deleuze and Guattari); the fabricated, which uses that concept to trace the processes through which the world has already been achieved (ANT, especially in its latest Latourian take); and the response-able, which sees assemblage thinking as a tool informing the political intervention on the assemblage that is created (Haraway). In other words, if 'Deleuze and Guattari ask "what is possible?", actor network theory [. . .] asks "what is required?"' and Haraway asks how one can '"become worldly and respond" to world's complexity' (Greenhough, 2011, p. 135). These different takes on assemblage thinking are not incompatible but form the evolving, experimental, and patch-worked vitalist sensibility that we embrace with our book (Harrison, Pile, and Thrift, 2004). Indeed, despite their differences these ways of approaching assemblage thinking share a morphogenetic view on reality, seeing it as a set of dynamic non-linear processes of human and non-human relations that lead to 'the production of life forms' and their related problematization and potentialities (Escobar, 2007, p. 107). The geographical scholarship emerging from this school of thought has fruitfully informed, to cite just a few, the analysis of world city topologies (Smith, 2003), the inter-scalar becoming of social movements (McFarlane, 2009; Davies, 2012; Vasudevan, 2014), policy transfer (McCann and Ward, 2013) and geopolitics (Dittmer, 2013), the codification of consumption (Dodge and Kitchin, 2005), and understandings of feminism and the body (Colls, 2012), as well as fundamentally rewriting our epistemological understanding of the urban sphere (Farías and Bender, 2010; McFarlane, 2011b; Amin, 2015; and, although they do not explicitly buy into the assemblage metaphor, Amin and Thrift, 2002). Yet, assemblage thinking has still to prominently enter the geographical study of marginality and bordering.

An exception has been the work of Colin McFarlane on Mumbai. McFarlane investigates how the city is learned through an incremental process of assembling with the urban machine and in doing so he reveals a world made of 'unequal relations of knowledge, practice and power' interlinked one to another and assembled with the post-human fabric of the urban (McFarlane, 2011b, p. 176). This world is explored in its becoming-urban, becoming-learned, and becoming-marginal. It is a world traced in its unfolding, in its being trans-local, not fixed, stretched in space and time. The insights given by this perspective are far from being only theoretical. They inform practical matters related to education, public space, and, most importantly, they offer reasoning around contextual political change (McFarlane and Rutherford, 2008; Desai, McFarlane, and Graham, 2014). On the same line we find the vivid ethnographic account provided by AbdouMaliq Simone, who investigates the urban margins to show the provisional assemblage of the urban economy (Simone, 2011), portraying the potentiality of life in the contemporary city of the South (Simone, 2010); the work of Ash Amin on the racialized other (Amin, 2010), encounter with difference (Amin, 2012), and on urban infrastructures and the 'slums' (Amin, 2014); the theoretical contributions of Dovey, who speaks of informal urbanism as a 'complex adaptive assemblage' (Dovey, 2012); and a few other scholars interested in development studies, homelessness, and race (Saldanha, 2006; Swanton, 2010; Lancione, 2013, 2014; Grove and Pugh, 2015).

In *Rethinking Life at the Margins* we build on this scholarship and also on the structural, relational, and post-colonial analytics described at the beginning of this chapter. Although the structural analytic is possibly the farthest from assemblage thinking, in this book we do not reject the existence of 'structures' *per se*, but we understand them as relationally built, always contested, and not dependent upon any specific variable or force. They are, as Chattopadhyay has put it, an 'infra-structure' territorialized material and immaterial relationships between human and non-human, flows and ideas, emotions and power, which contribute to the definition of a particular course of action (Chattopadhyay, 2012). From the relational and grounded analytic, we take the need for a direct engagement with the spaces and people of the margins. Taking inspiration from the best ethnographic work available, our aim is to link the grounded dynamics of the margins to their structural elongations, without aiming to explain these relationships on the basis of pre-established knowledge or social categories. Rather – and this is what we draw from the post-colonial analytic – established knowledge is criticized and challenged through the production of alternative knowledge aiming at the constitution of a differential cultural politics of the margins.

To summarize, the principal traits of the approach developed in this book can be listed as follows:

A) We propose a vitalist perspective on the margins, deployed through assemblage thinking. The latter is the preferred approach to field research and theory construction, as it rejects over-arching theories of the margins, of their spaces and subjects. Rather, it invites a grounded experimental encounter open to the multiplicity of life and to its differential potential of becoming;
B) We promote a processual approach, interested in the unfolding of action, in tracing how things are assembled and in the thick description of their articulations (what Deleuze and Guattari call territorialization – deterritorialization – reterritorialization);
C) In the articulation of life, human and non-human are on the same ontological plane: there is no prominence of one or the other. A clear reassessment of non-human agencies in the production of life at the margins is a central point of rethinking marginality through assemblage;
D) The scope of a critical approach to assemblage thinking is to take the potential of *a life* back into the representation of life at the margins, in order to assess how different articulations may be possible and to highlight what currently impedes them;
E) This is a critical approach because its object is about the search for an affirmative and liberating cultural politics of the margins. Coherently with its immanent stances, it does not provide a once-and-for-all recipe for political action, but it invites the cultivation of a minor politics, where 'minor' has to be interpreted in a Deleuzian sense as a politics characterized by its capacity to become, to articulate difference, and to be open to new articulations that are not to be subsumed by 'a redundant majority' (Conley, 2010, p. 167).

On this book: context, subjects, and politics

The above points enlighten a path but do not establish a strict route. Reapproaching marginality through assemblage thinking means first and foremost to experiment and to explore, and it requires a high degree of theoretical and empirical eclecticism (Rankin, 2011). None of the authors presented in this book would consider themselves to be strictly a 'Deleuzian', and some of the chapters in the book do not address the notion of 'assemblage' explicitly, nor refer to one specific post-human philosophy. This lack of explicit reference or over-theorization *has been a conscious choice.* The idea of the book is to provide an accessible and heterogeneous collection of essays that, although sharing the same ethos summarized by this introduction, articulate it with due difference and respect to contextual circumstances and cultural traditions.

In what follows I expand on the specific theoretical foundation necessary to reapproach contexts, subjects, and politics, and I also present the contributions making up this volume. Although the chapters are organized into three parts, contexts, subjects, and politics are obviously only analytically separated. Each of the chapters touch upon all of these elements, sometimes stressing one more than the other for the sake of argument and clarity. Dividing them is simply a strategy to show the possibilities offered by the proposed approach.

Recontextualization

The first necessary step is to promote a vitalist understanding of the contexts of marginalization. This is necessary in order to trace and represent the heterogeneity of the components, processes, and forces that make up life at the margins. One promising resource towards this end is a specific 'branch' of assemblage thinking, namely the literature on urban assemblages, which sits at the intersection of Deleuzo-Guattarian thinking and Actor Network approaches (Farías and Bender, 2010; McFarlane, 2011a). Assemblage urbanism conceives the city as a mechanosphere of vitalist entanglements always open to possibility and change (Amin and Thrift, 2002). If one takes these articulations seriously, the city cannot be a place to which we – as humans – react by means of practicing it but as a co-constituted space of what Thrift calls 'outstincts' (as opposed to the more deterministic 'instincts'; Thrift, 2014). The city – like any other context – is *alive*, and the marginal subject is alive *through* and *with* it. One could say, in this sense, that the urban comes for the first time to be seen precisely as a 'web of life', in which human and non-human 'are bound together in a vast system of interlinked and interdependent lives' (Park, 1936, p. 1). The difference from the traditional account of urban ecology is, however, vast. It is no longer a matter of understanding urbanism as a way of life – hence as a 'way of behaving' linked to a typically urban way of 'thinking and conduct' (Anderson, 1959, p. 1) – but of understanding the many lives that make up the urban. The ecology is not read through human eyes and it is not human dependent but emerges from the entanglements that make up the world and that *may make up the world.* This is, in other words, an echo-logic: a constant refrain of heterogeneous machinic affiliations (Guattari, 2009), a productive reverberation of the multiple forms of lives taking place in the urban (what Farías calls the 'ecological process' of the urban; 2011, p. 368).

Drawing from this take on the urban, we understand marginal contexts both as relational spaces and as a rhizome of eventful post-human 'crossroads' (Simone, 2010) within which multifaceted experiences of the margins are constituted. The margins cease to be just a matter of places ('slums', sidewalks, waste-fields, informal camps), services (soup kitchens, shelters, drop-ins), institutions and procedures (asylum, social housing, and minimum wage bureaucracies), and so on. Rather, they become a matter of entanglements between objects and bodies, discourses and power, performances and blueprints for action – a universe of capacities that need to be traced in their contextual deployment in space and time. The task of the critical assemblage thinker is to recontextualize marginalization through the careful analysis of the multiple capacities making up the contexts of the margins. Contexts are never easily depicted and borders are always shifting, overlapping, made up of matter that resembles a patchwork more than a neat cartography of exclusion (Mezzadra and Neilson, 2013). Reapproaching the context of marginalization means to understand that these contexts are 'founded on the blurring of boundaries rather than their repeated delineation' (Gandy, 2005, p. 29). It is within that vitalist mingling and blurring that a micro-politics of reverberation, an echo-logic of dispossession and expulsion but also of affirmation and capacitation take place, and it is there – within those post-human entanglements – that marginality needs to be traced and understood.

The first four chapters of this book illustrate this point. They reimagine their specific marginalized context with the aim of highlighting things and processes previously unseen or not fully put into focus or problematized. These chapters offer insights into how to rewrite our understandings of the visions, architectures, public artefacts, and infrastructures through which borders and margins are lived and performed. Kavita Ramakrishnan reads a resettlement colony in Delhi through the urban aspirations of its people, showing how the 'centre of Delhi' acts as a vitalist catalyst of dreams, visions, and aspirations affecting how people live and perform their marginal condition. What Ramakrishnan does is to recontextualize mainstream readings of peripheral resettlement colonies via bringing the 'city' – as assemblage of design, practices, and vernaculars – into the relational constitution of the margins. The most important insight that we gain from this chapter comes from the voices of Ramakrishnan's informants. They are the ones who, from the margins, are recontextualizing the city, and it is only taking their recontextualization seriously that one may be able to reassess marginals' political claims.

In Francesca Governa and Matteo Puttilli's chapter the often overlooked power of urban architectures and public spaces is brought into focus to describe the changes underway in post-revolutionary Tunis. In their account the city is fully into the scene: streets, walls, buildings, fences, squares, and more are not seen as a platform of the revolution, but as co-constitutive elements of old and new practices, habits, and use. Recontextualization here operates at two levels. The first is the level of what Tunis – as a mechanosphere of people and matter – *has done* in reconfiguring its political grounding. The second is the description of this reconfiguration by Governa and Puttilli, which is done through the lens of the built city showing the centrality and hybridity of space in the (re)doing of marginalization. Another important layer to this contribution is methodological, since the chapter originates from an extensive video engagement with Tunis, which may offer further insights into how to reconceptualize the contexts of the margins.[1] Moving further 'down', into the granularity of spatial practices and vitalist entanglements, Mark Tirpak's chapter on the ethnography of food trucks in San Antonio (Texas) recontextualizes the marginalization of public spaces through the thick description of small assemblages of power. What we see here is the grounded and vitalist rewriting of the bordering practice of gentrification through dog fences, contradictory regulations, and expensive tacos. What Tirpak offers us is a way of looking at an over-studied process of bordering (gentrification) through a fresh, grounded vitalist lens, which achieves two important results. First, it allows him and us to understand where, when and how marginality is assembled on the ground – pointing to specific devices and designs that border things in and out. Second, it clearly connects these micro-assemblages to established (molar, in a Deleuzian language) patterns of exclusion in US cities.

If Ramakrishnan, Governa, Puttilli, and Tirpak render the contexts of the margins alive through its visions, spaces, and micro-assemblages, then AbdouMaliq Simone's chapter concludes this section showing that these and other things are not exceptional – but they are *just* what the city is: a complex repository of human and non-human agencies entangled one into the other. Within these entanglements injustices and borders are made and remade continuously, escaping any over-arching understanding, definition, or politics. What instead is needed is an appreciation of the power enmeshed in the day-by-day doing of city making, in the repairing and remaking of relations between inhabitants and places. It is there, within the unfolding of those assemblages, that according to Simone a 'just' city could be imagined and crafted: one able to build upon the repository of experiences, opportunities, and vitalist energies continuously making and remaking life in the urban. What this chapter and the previous show us is the main point of this first part of the book: in order to reimagine the margins politically one has to reimagine how things assemble in contexts. It is only in recontextualizing the margins that new grounds for equitable ends could again be seen and argued for.

Resubjectification

To reapproach the margins from a vitalist perspective human subjectivity does need to be rethought alongside context: it has to be opened up to its wider environment, to acknowledge its hybrid constituency. In this sense the human subject should no longer be seen as the 'disengaged first-person-singular self [. . .] self-reliant for her or his judgements on life, the universe and everything' (Pile and Thrift, 1995, p. 14), but as an assemblage made up of many different assemblages shaped by countless 'lively' encounters. The marginal subject does not only live *in* context, but s/he factually becomes *with* and *through* the mechanosphere of that context. In order to navigate this intricateness one needs a non-reductive, non-Cartesian account of subjectivity. Examples can be found in the work of diverse philosophical thinkers, such as Lefebvre, Derrida, Butler, and Foucault, who move away from the Freudian and Lacanian subject – a subject locked into phallocentrism and textuality – in order to account for the variegated 'contextuality' of the self (Wylie, 2010). With Deleuze and Guattari that 'contextuality' is fully rendered visible and its complexity is taken into account: the subject ceases to be only matter of text, body, rationality, or power relations, but becomes understood as an ongoing, never-finished assemblage of the human, non-human, discursive, technical, and potential matters making up the world (Deleuze and Guattari, 1987).

The subject is just one of the parts (breakable into smaller parts) of the contexts described above. It is defined by it, it is made by it, it is constituted through it, and it does constitute it: the subject is not determined by the strict boundaries of the rational self because rationality itself is just a product (and a producer) of countless other machines (Deleuze and Guattari, 1977). Deleuze, and more particularly Guattari (Guattari, 2010, 2013), do not bring to the fore a definitive theory of the subject but allow us to navigate the broad rhizomatic canvas where the subject becomes with the other assemblages of the world (Lancione, 2016). The key point is to understand the (marginal) subject as being 'collective': meaning by this that the subject is the expression of 'the heterogeneity of the components converging to produce subjectivity' (Guattari, 1996, p. 193). This does not mean that the subject disappears in a collective mist, but simply that in order to grasp the subject one needs to look at the collective process of machinic affiliation through which subjectivity is constituted, challenged, and reassembled. This means to understand the inner-self as an elongation of the mechanosphere and vice-versa, in their productive constitutive tensions. Rosi Braidotti has depicted this process very clearly:

> [S]ubjectivity is a socially mediated process of entitlements to and negotiations with power relations. Consequently, the formation and emergence of new social subjects is always a collective enterprise, 'external' to the individual self while also mobilising the self's in-depth and singular structures.
> (Braidotti, 2011, p. 18)

The shift from canonical scholarship on marginalization is considerable. The experience of marginalized people cannot be understood any longer as a matter of personal culpability or lack of will, nor simply as the outcome of broader economic causes, but can only be grasped as an ongoing process of subject formation where the latter is always a collective endeavour – a matter of contextual material and discursive arrangements to be addressed in their heterogeneous becoming. In approaching the marginal subject from this perspective we are looking for performances of a more-than-human kind: pre-cognitive and cognitive; rational and machinic; emotional and affective; diagrammatic and enabling. This is a contextual endeavour because it is made in context, through context.

Think for instance of an hypothetical inhabitant of a 'slum'. Her subjectivity is constituted at the level of day-to-day operation with the multiplicities making up life in the 'slum'. She is related to the materiality of the place; its internal rules and power dynamics; the encounter with foreign aid; the law; traditions; social customs; the lack of facilities and the opportunities offered by what is at hand; affects and fears; the aesthetics of life in the overcrowded 'shanty' town; plastic bags filled with water; wood; petrol; recycled tins; sunshine; ugali[2]; rain; mosquitoes; etc. These machinic-assemblages operate, as Guattari would have it, at 'the heart of human subjectivity' (Guattari, 2009): they mould desires, constantly reconfiguring *how* the subject is – in becoming. The subject has a voice, so a minimal humanism is still there (Thrift, 2008). However, this voice is not individualized and cannot be understood as such: the voice is a chorus (a refrain, an echo-logic) made up of all the above entanglements. In short, from a vitalist perspective the marginalized subject cannot be separate from the wider social machine that co-constitute 'her' subjectivity: from the influence of external capital in shaping the price of staple foods in the 'slum', to informal economies of subsistence; from the role of the law, to the relevance of ghosts, traditions, and customs; from the galvanized steel of the roof under which she sleeps to the social-aid campaign advertised with the face of George Clooney. The challenge of instantiating a vitalist take on the margins is to describe this complex machinic subject without reducing its dramatic complexity.

Tawhanga Mary-Legs Nopera's account of the Maori subject in New Zealand fully embraces this collective notion of subjectivity. Far from being 'simply' post-colonial in its claims, Nopera's chapter shows us how it is possible to rethink one's own marginality via the reconsideration of what makes up one's own subjectivity. Importantly, Nopera does so via reappropriating Deleuze and Guattari's thought through traditional Maori knowledge – namely through the notions/praxis of *whakapapa* and *raranga*. The subject that emerges in this account is a machine-of-machines, assessed in its own contextual (cultural and political) terms. This is a subject that is hard to distinguish from its context because the two are just in-becoming, constantly rewriting their marginality, in a mist of traumas, powerful affects, and displacements – as Nopera's chapter so beautifully shows.

The following two chapters, by Tatiana Thieme and Gaja Maestri, are united by the same effort of reapproaching the subject through detailed accounts of how life at the margins constantly rewrites itself. Thieme offers a rich ethnography of the changing aspirations and internal conflict that young 'hustlers' living in Mathare, Nairobi, face in their struggles to improve their life conditions. Thieme's work is able to show the granularity of the resubjectification process through the story of a young woman, Eliza, who passes from the struggle to affirm her identity in the male-centred waste economy, to having to balance her aspiration to do 'well' with her belonging to the 'hood'. The struggles of Eliza are traced as assemblages constructed with and through the wider 'mechanosphere' of the 'hood', thus offering a rich ethnographic example of how it may be possible to rewrite subjectivity following a non-normative approach. In a similar vein, Maestri's work on Roma people in Rome, Italy, powerfully shows how Roma people themselves – again, in their struggle for dignity and affirmation – are able to rewrite their own subjectivity. Pivotal in this process is the assemblage that these people make with various urban machineries – policies, police interventions, abandoned factories, etc. – which eventually brings them from being defined, and perceived, as 'Roma' to becoming 'squatters'. According to Maestri, it is thanks to assemblage thinking that scholars can trace these processes of resubjectification, giving them theoretical (and political) resonance, as her chapter clearly does.

Last but not least, Jean-Baptiste Lanne brings us back to Nairobi, this time to the 'slum' of Kibera. His aim is not to show processes of resubjectification of the people he looked at, but to question the process of resubjectification that we – as researchers – bring about in reapproaching life at the margins. According to Lanne, we need to marginalize ourselves in order to recentre the margins that we aim to investigate. It is only by doing so that a true vitalist ethnography of the margins can be brought to the fore, as the powerful 'recordings' Lanne presents to us clearly show. His chapter brings into focus the underlying issues of positionality and methodology that the previous (and following) chapters have implicitly dealt with, offering therefore a valuable platform from which to think about the concrete 'doings' of rethinking life at the margins and of redoing subjectivity from a vitalist perspective.

Repoliticization

The previous chapters do not limit themselves to contexts and subjects but are deeply connected to the third part of the book, which is about repoliticization. The acts of recontextualizing and resubjectifying the margins are, indeed, related to the particular cultural politics that we aim to promote with this book. There are two main points making up the ethics fuelling this politics, which should also render clearer why we have decided to focus the first two parts of the book on rethinking 'contexts' and 'subjects'.

First, a post-human ethics on the margins is always contextually driven. Without diminishing the importance of basic fundamental and universal rights, it is not in the light of these rights that we are moved politically, but in their contextually based (de)construction. If the former are usually deployed around human-centred meta-narratives of what is considered to be good or bad, the latter mean to reapproach these universals from the day-by-day enmeshment of human and non-human alike (Braidotti, 2013).[3] In other words, the political relevance of rethinking life at the margins is found within, and not outside, the particular context of action. This is coherent with the overall Deleuzian-Guattarian ethical project, centred around the notion of affects – the capacity of something to affect something else (Colebrook, 2004). In short: 'Because affective relations are variable and contextual, and power is simultaneously a force of composition and of decomposition affecting participating bodies at multiple sites of engagement, *the "rightful" limits to the impact of one body's powers upon another can only be decided according to the context of the situation in which their meeting occurs*' (Bignall, 2010, pp. 96–97, emphasis added).

Second, a post-human ethics of the margins is about a vitalist take on life and its subjects. Preserving and raising human life is not a matter of human life itself, or defined by it, but on the recognition that *life* is only *one* and it is shared by human and non-human alike (as, indeed, 'a' plane, or common ground). Bennett has, in this sense, called for a vital materialism that is able to 'raise the status of the materiality of which we are composed' (Bennett, 2010, p. 12) in order to give it and thus give us, together, prominence and value (on a similar vein, see Stengers's notion of 'cosmopolitics'; Stengers, 2005). In this sense, a post-human ethics of the margins is about what Grosz has called 'partial objects': 'organs, processes, flows, which show no respect for the autonomy of the subject. *Ethics is the sphere of judgements regarding the possibilities, and actuality of connections, arrangements, linkages, machines*' (Grosz, 1993, p. 172, emphasis added).

To summarize and to clarify, killing, violating, exploiting, and all sorts of actions that do not allow for the full and positive expression of all life are to be considered unacceptable *a priori*. But one should not stop at this assessment. Rather, we are interested in showing how the positive and negative articulations of life take place in the contextually based post-human entanglements making up the margins. If humans are just 'dividuals' – 'who for the most of the time are simply part of a combination of bodies or parts of bodies, resonating around a particular matter of concern' (Amin and Thrift, 2013, p. 50) – for us acting politically is not about building universalizing knowledge and grand narratives of the good and the bad, but about tracing that 'combination of bodies or parts of bodies' and revealing the nuances of power in its making. It is about learning how to 'inhabit' the struggles of the margins and then 'find new languages and registers with which to name and possibly displace it' (Pieterse, 2013, p. 29). In doing so, we work towards an unfinished and shifting alternative cultural politics and cultural geography of marginality. The four chapters comprising this third part of the book – like the eight that preceded – are examples of the numerous ways through which rethinking life at the margins entails politics. Their core aim is not, once again, to provide a theory of what is 'political' at the margins, but to offer various examples of how, tracing and rendering visible the 'partial objects' making up life at the margins, one is able to reveal new political stakes as well.

This part of the book begins with Francisco Calafate-Faria's contribution, which takes us amongst Brazilian waste pickers in order to substantially challenge canonical scholarship on the matter. Besides the production of alternative knowledge, Calafate-Faria shows us the potential for a counter-politics of waste that starts from an understanding of the post-human constitutive forces at play in movements of recycling and countercycling. Giving full agency to rubbish – and differentiating this agency according to different materials, contexts, and subjects – Calafate-Faria properly shows what critical assemblage thinking is: a quest to trace life in its own place, in order to show where and how it could be rearticulated. In a similarly grounded and embodied way, in the following chapter Eszter Krasznai Kovács provides us with a powerful ethnographic account of marginalization processes in rural Hungary. It is only thanks to her careful tracing of how humans' claims and desires assemble with 'nature' and the EU's policies that Krasznai Kovács is able to show the 'out-of-place'-ness of Hungarian and EU laws and rules. This is an 'out-of-place'-ness made political through a grounded recontextualization of what is considered to be urban and what is considered to be rural in contemporary Hungary, *and* a reassessment of how marginalized subjects are produced within the powerful diagrams activated around normative understandings of rural.

Again through long-term ethnographic fieldwork and engagement, in the next chapter Elisabetta Rosa brings us within the day-to-day practices of Roma people living in Turin, Italy. Rosa's account is political from the outset: she aims to reframe marginality as a 'resource', in order precisely to reconstruct that oppositional value-set to which hooks was referring to (1994). To do so, Rosa shows us the constitutive mingling of bodies and urban machines through which being marginalized can be rethought in terms of unexpected capacities, reappropriations, alternative cartographies, and powerful non-human agencies. The margins, according to Rosa's reading, are not opposite to the centre – an understanding that reproduces normative paradigms and policies – but a repository of practice-based resources that need to be assessed from within.

If the previous chapter has shown us how the vitalist analytic can render things political and can show the nuance of politics at the margins, then the last chapter in this part adds another important touch to this picture. Cheryl Gilge brings us to an apparent virtual context: that of web 2.0. She shows the potential of assemblage thinking to identify political issues where apparently there are none, within grounds where it seems that there are only gains. To do so, Gilge carefully analyzes the disturbing way through which open source and the web 2.0 marginalizes labour, carefully revealing how this process is a matter of *microfascism*. Her contribution shows how thinking through assemblage can let one see – describe and understand – the production of contemporary ubiquitous borders, showing also the potential of critique in challenging them. Gilge – like Calafate-Faria, Krasznai Kovács, and Rosa – *does* the politics-of-the-assemblage: the unfinished work of showing the mingling of differential agencies within which alternative lines of flight can be enacted.

Minor thinking

The best way for this book to go further is to question itself and ourselves. That's why the book has no conclusion, but a self-critical opening. In it, Darren J. Patrick poignantly challenges all of us by asking: Why do we bother speaking of the 'assemblage of life at the margins' in the first place? What is that for? Whom is it for? And on what grounds do we do it? Patrick does the fundamental job of bringing 'assemblage' back to where it came from, and in doing so he discusses the premises upon which this book is built. Through this account, Patrick implicitly reassesses an important point: that rethinking life at the margin is about sharing an ethos and is neither about building (or looking forward to) a definite statement around marginality, nor about establishing 'assemblage thinking' as *the* approach to enquire the margins. The book is about challenging those statements in the first place and, as Patrick shows us, to challenge our own positions, institutions, and established knowledge too.

To build this book we have drawn implicitly and explicitly from established analytics of the margins and emergent vitalist ontologies. Marginal contexts and subjects are understood as a set of post-human articulations that are always open to the emergence of new events. This is not about romanticizing life at the margins, but about widening the scope of what we look for when we approach that life in the first place. The task of reimagining it is as urgent as it is messy, but it is also possible, as the authors of this book show. The key is to be ready to accept the intrinsic multiplicity of life itself. And here lies the contribution of this book: it proposes a concrete way to begin the necessary work in order to acknowledge and be able to investigate the multiplicity of life. It does so through a vitalist lens, applying assemblage thinking as a tool to reapproach marginal contexts and subjects. These are reframed to more precisely consider the entanglements of bodies, matter, knowledge, power, affects, and so on, always (re)making margins and borders. Moreover, recontextualization and resubjectification are not academic exercises but must lead to repoliticization: of matter, of bodies, of borders – of *a life* at the margins.

As editor of this volume, I'd like to think of this as a micro-political effort, a minor theory crafted to push our understanding of the margins and their borders a bit further (Deleuze and Guattari, 1986; Katz, 1996). 'Rethinking' does not stand as a finalized action. We have not 'rethought' marginality with this volume. What we have done is provide some solid case studies around how it may be possible to look at the contexts, subjects, and politics of the margins differently, in a way that may invite people towards the performative act of constantly 'rethinking' what is taken for granted around marginality. We hope that this minor thinking may allow readers to articulate new political questions and strategies. The book offers insights in this sense from at least two levels. The first is at the level of each singular chapter, with its own contextually based and relevant insights on specific cases of marginalization offering a wide array of rich ethnographies written from the North, the South, the body, the square, the park, the rural, the virtual, and much more – an heterogeneity that is there to connect, to be used, to be multiplied. The other is at the level of the collective subjectivity enacted by these chapters: a tool, an assemblage, the sketch of a cultural politics that we hope people will criticize, take apart, and bring forward to enhance alternative understandings of life and its margins.

Acknowledgements

Thanks to Ash Amin, Eszter Krasznai Kovács, Kavita Ramakrishnan, Cheryl Gilge, and Eleonora Mignoli for their edits and comments on this chapter.

Notes

1 See the web-documentary 'Al centro di Tunisi', which can be accessed at the following page: http://webdoc.unica.it/ (In French and Italian).
2 Stiff cornmeal porridge, a common staple food in east Africa.

3 It is worth stressing that 'context' does not equate by any means with 'local'. A context is 'a plural event which is more or less spatially extensive and more or less temporally specific. It is, in other words, a parcel of socially constructed time-space which is more or less elongated' (Thrift, 1996, p. 43), which essentially means that a context might be stretched along considerable relational distance (Latour, 2005), as is usually the case at the margins.

References

Amin, A., 2010. The remainders of race. *Theory, Culture & Society*, 27(1), pp. 1–23.

Amin, A., 2012. *Land of strangers*. Cambridge: Polity.

Amin, A., 2013. Telescopic urbanism and the poor. *City*, 17(4), pp. 476–492.

Amin, A., 2014. Lively infrastructure. *Theory, Culture & Society*, 31(7/8), pp. 137–161.

Amin, A., 2015. Animated space. *Public Culture*, 27(2), pp. 239–258.

Amin, A. and Thrift, N., 2002. *Cities, re-imaging the urban*. Cambridge: Polity.

Amin, A. and Thrift, N., 2013. *Arts of the political. New openings for the Left*. London: Duke University Press.

Anderson, B., 2009. Affective atmospheres. *Emotion, Society and Space*, 2(2), pp. 71–81.

Anderson, B. and Harrison, P., 2010. The promise of non-representational theories. In: B. Anderson and P. Harrison, eds., *Taking-place: non representational theories and geography*. Farnham: Ashgate, pp. 1–36.

Anderson, N., 1959. Urbanism as a way of life. In: N. Andersen, ed., *The urban community: a world perspective*. New York: Holt, Rinehart & Winston.

Arabindoo, P., 2011. Rhetoric of the 'slum'. *City*, 15(6), pp. 636–646.

Ashcroft, B., Griffiths, G. and Tiffin, H., 2007. *Post-colonial studies: the key concepts*. London: Routledge.

Barnett, C., 2006. Postcolonialism: space, textuality and power. In: S. Aitken and G. Valentine, eds., *Approaches to Human Geography*. London: Sage, pp. 147–159.

Bennett, J., 2010. *Vibrant matter. A political ecology of things*. Durham: Duke University Press.

Besio, K. and Butz, D., 2004. Autoethnography: a limited endorsement. *The Professional Geographer*, 53(March 2015), pp. 432–438.

Bignall, S., 2010. Affective assemblages: ethics beyond enjoyment. In: S. Bignall and P. Patton, eds., *Deleuze and the postcolonial*. Edinburgh: Edinburgh University Press, pp. 78–103.

Billson, J.M., 1996. No owner of soil: redefining the concept of marginality. In: D. Routledge, ed., *Marginality: Theoretical perspectives*. Thousand Oaks: Sage, pp. 29–47.

Bradatan, C., and Craiutu, A., 2012. "Introduction: The Paradoxes of Marginality." *European Legacy*, 17(6), pp. 721–729.

Bondi, L., 2005. Making connections and thinking through emotions: between geography and psychotherapy. *Transactions of the Institute of British Geographers*, 30, pp. 433–448.

Bourgois, P. and Schonberg, J., 2009. *Righteous dopefiend*. Los Angeles: University of California Press.

Braidotti, R., 2011. *Nomadic subjects. Second edition*. New York: Columbia University Press.

Braidotti, R., 2013. *The posthuman*. London: Polity.

Castells, M., 1977. *The urban question: a Marxist approach*. Cambridge: MIT Press.

Chattopadhyay, S., 2012. *Unlearning the city. Infrastructure in a new optical field*. Minneapolis: University of Minnesota Press.

Clarke, J. and Martin, C., 2004. *Anarchy, geography, modernity*. Lenham, MD: Lexington Books.

Colebrook, C., 2004. The sense of space: on the specificity of affect in Deleuze and Guattari. *Postmodern Culture*, 15(1), pp. 1053–1920.

Colls, R., 2012. Feminism, bodily difference and non-representational geographies. *Transaction of the Institute of British Geographers*, 37(3), pp. 430–445.

Conley, V., 2010. Minoritarian. In: A. Parr, ed., *The Deleuze Dictionary. Revised Edition*. Edinburgh: Edinburgh University Press, pp. 166–168.

Darling, J., 2011. Domopolitics, governmentality and the regulation of asylum accommodation. *Political Geography*, 30(5), pp. 263–271.

Datta, A., 2012. *The illegal city. Space, law and gender in a Delhi squatter settlement*. Farnham: Ashgate.

Davies, A.D., 2012. Assemblage and social movements: Tibet support groups and the spatialities of political organisation. *Transactions of the Institute of British Geographers*, 37(2), pp. 273–286.

Davis, M., 2006. *Planet of slums*. New York: Verso.

DeLanda, M., 2005. Space: extensive and intensive, actual and virtual. In: I. Buchanan and G. Lambert, eds., *Deleuze and space*. Edinburgh: Edinburgh University Press, pp. 80–88.

Deleuze, G., 2001. *Pure immanence*. New York: Zone Books.

Deleuze, G. and Guattari, F., 1977. *Anti Oedipus. Capitalism and schizophrenia*. New York: Penguin Books.

Deleuze, G. and Guattari, F., 1986. *Kafka. Toward a minor literature*. Minneapolis: University of Minnesota Press.

Deleuze, G. and Guattari, F., 1987. *A thousand plateaus*. New York: Continuum.

Del Pilar, J.A. and Udasco, J.O., 2004. Marginality theory: the lack of construct validity. *Hispanic Journal of Behavioral Sciences*, 26(1), pp. 3–15.

Desai, R., McFarlane, C. and Graham, S., 2014. The politics of open defecation: informality, body, and infrastructure in Mumbai. *Antipode*, 47(1), pp. 98–120.

Desjarlais, R., 1997. *Shelter blues: sanity and selfhood among the homeless*. Philadelphia: University of Pennsylvania Press.

Dewsbury, J.-D., 2000. Performativity and the event: enacting a philosophy of difference. *Environment and Planning D: Society and Space*, 18, pp. 473–469.

Dittmer, J., 2013. Geopolitical assemblages and complexity. *Progress in Human Geography*, 38(3), pp. 385–401.

Dodge, M. and Kitchin, R., 2005. Codes of life: identification codes and the machine-readable world. *Environment and Planning D: Society and Space*, 23(6), pp. 851–881.

Dovey, K., 2012. Informal urbanism and complex adaptive assemblage. *International Development Planning Review*, 34(4), pp. 349–368.

Duneier, M., 1999. *Sidewalk*. New York: Farrar, Straus and Giroux.

Escobar, A., 2007. The 'ontological turn' in social theory. A commentary on 'human geography without scale', by Sallie Marston, John Paul Jones II and Keith Woodward. *Transactions of the Institute of British Geographers*, 32(1), pp. 106–111.

Farías, I., 2011. The politics of urban assemblages. *City*, 15(3–4), pp. 365–374.

Farías, I. and Bender, T. eds., 2010. *Urban assemblages: how actor-network theory changes urban studies*. London: Routledge.

Ferguson, R., Gever, M., Minh-ha, T.T. and West, C. eds., 1990. *Out there. Marginalization and contemporary culture*. New York: The New Museum of Contemporary Art.

Foucault, M., 2000. *Power. Essential works of Foucault, 1954–1984, Volume 3*. London: Penguin.

Gandy, M., 2005. Cyborg urbanization: complexity and monstrosity in the contemporary city. *International Journal of Urban and Regional Research*, 29(1), pp. 26–49.

Gibson-Graham, J.K., 1994. 'Stuffed if I know!': Reflections on post-modern feminist social research. *Gender, Place & Culture*, 1(March 2015), pp. 205–224.

Gowan, T., 2010. *Hobos, hustlers and back-sliders: homeless in San Francisco*. Minneapolis: University of Minnesota Press.

Greenhough, B., 2011. Assembling an island laboratory. *Area*, 43(2), pp. 134–138.

Grosz, E., 1993. A thousand tiny sexes: feminism and rhizomatics. *Topoi*, 12(2), pp. 167–179.

Grove, K. and Pugh, J., 2015. Assemblage thinking and participatory development: potentiality, ethics, biopolitics. *Geography Compass*, 9(1), pp. 1–13.

Guattari, F., 1996. Subjectivities. For the better and for the worse. In: G. Genosko, ed., *The Guattari reader*. Oxford: Blackwell Publishers Ltd, pp. 193–203.

Guattari, F., 2009. *Chaosophy. Texts and interviews 1972–1977*. Los Angeles: Semiotext(e).

Guattari, F., 2010. *The Machinic unconscious: essays in schizoanalysis*. London: Semiotext(e).

Guattari, F., 2013. *Schizoanalytic cartographies*. London: Bloomsbury.

Gunew, S.M., 1994. *Framing marginality. Multicultural literary studies*. Melbourne: Melbourne University Press.

Hall, S., 1992. The West and the rest: discourse and power. In: S. Hall and B. Gieben, eds., *Formations of modernity*. London: Polity, pp. 275–331.

Haraway, D.J., 1991. *Simian, cyborgs, and women. The reinvention of nature*. London: Free Association Books.

Harrison, S., Pile, S. and Thrift, N., 2004. *Patterned ground*. London: Reaktion Books.

Harvey, D., 1973. *Social justice and the city*. London: Edward Arnold.

hooks, b., 1989. Choosing the margin as a space of radical openness. In: *Yearnings: race, gender and cultural politics*. New York: South End Press, pp. 203–209.

hooks, b., 1994. *Outlaw culture*. New York: Routledge.

Huggan, G., 2001. *The postcolonial exotic*. New York: Routledge.

Jacobs, J.M., 2012. Urban geographies I: still thinking cities relationally. *Progress in Human Geography*, 36(3), pp. 412–422.

Johnsen, S., Cloke, P. and May, J., 2005. Transitory spaces of care: serving homeless people on the street. *Health & Place*, 11(4), pp. 323–336.

Katz, C., 1994. Playing the field: Questions of fieldwork in geography. *The Professional Geographer*, 46(1), pp. 67–72.

Katz, C., 1996. Towards minor theory. *Environment and Planning D: Society and Space*, 14, pp. 487–499.

Katz, C., 2001. On the grounds of globalization: a topography for feminist political engagement. *Sign*, 26(4), pp. 1213–1234.

Lancione, M., 2013. Homeless people and the city of abstract machines. Assemblage thinking and the performative approach to homelessness. *Area*, 45(3), pp. 358–364.

Lancione, M., 2014. Entanglements of faith: discourses, practices of care and homeless people in an Italian City of Saints. *Urban Studies*, 51(14), pp. 3062–3078.

Lancione, M., 2016. The city and 'the homeless': machinic subjects. In: H. Frichot, C. Gabrielsson and J. Metzger, eds., *Deleuze and the city*. Edinburgh: Edinburgh University Press, pp. 145–160.

Latour, B., 2001. Gabriel Tarde and the end of the social. In: P. Joyce, ed., *The social in question. New bearings in history and the social sciences.* London: Routledge, pp. 117–132.

Latour, B., 2005. *Reassembling the social.* Oxford: Oxford University Press.

Law, J., 2009. Actor network theory and material semiotics. In: B.S. Turner, ed., *The new Blackwell companion to social theory.* London: Blackwell Publishing, pp. 141–158.

Law, J. and Callon, M., 1992. The life and death of an aircraft: a network analysis of technical change. In: W. Bijer and J. Law, eds., *Shaping technology/building society: studies in sociotechnical change.* Cambridge: The MIT Press, pp. 21–52.

Massey, D., 1993. Power geometry and a progressive sense of place. In: J. Bird, B. Curtis, T. Putnam, G. Robertson and L. Tickner, eds., *Mapping the futures: local cultures, global change.* London: Routledge, pp. 59–69.

Massumi, B., 1992. *A user's guide to capitalism and schizophrenia. Deviations from Deleuze and Guattari.* Cambridge: The MIT Press.

McCann, E. and Ward, K., 2013. A multi-disciplinary approach to policy transfer research : geographies, assemblages, mobilities and mutations. *Policy Studies*, 34(1), pp. 2–18.

McFarlane, C., 2009. Translocal assemblages: space, power and social movements. *Geoforum*, 40(4), pp. 561–567.

McFarlane, C., 2011a. Encountering, describing and transforming urbanism. *City*, 15(6), pp. 731–739.

McFarlane, C., 2011b. *Learning the city: knowledge and translocal assemblage.* Oxford: Wiley Blackwell.

McFarlane, C. and Rutherford, J., 2008. Political infrastructures: governing and experiencing the fabric of the city. *International Journal of Urban and Regional Research*, 32(June), pp. 363–374.

Mele, C., 2013. Neoliberalism, race and the redefining of urban redevelopment. *International Journal of Urban and Regional Research*, 37(2), pp. 598–617.

Mezzadra, S. and Neilson, B., 2013. *Border as method, or, the multiplication of labor.* Durham, NC: Duke University Press.

Mingione, E., 2004. Poverty and social exclusion in European cities: diversity and convergence at the local level. *City*, 8(3), pp. 381–389.

Mitchell, D., 1997. The annihilation of space by law: the roots and implications of anti-homeless laws in the United States. *Antipode*, 29(3), pp. 303–335.

Murdoch, J., 2006. *Post-structuralist geography.* London: Sage.

Nash, C., 2002. Cultural geography: postcolonial cultural geographies. *Progress in Human Geography*, 26, pp. 219–230.

Neuwirth, R., 2007. Squatters and the cities of tomorrow. *City*, 11(1), pp. 71–80.

Obeng-Odoom, F., 2013. Book review: arrival city: how the largest migration in history is reshaping our world. *African Review of Economics and Finance*, 5(1), pp. 76–78.

O'Connor, A., 2001. *Poverty knowledge: social science, social policy, and the poor in twentieth-century U.S. history.* Princeton: Princeton University Press.

Orlikowski, W.J., 2007. Sociomaterial practices: exploring technology at work. *Organization Studies*, 28(9), pp. 1435–1448.

Pain, R., 2003. Social geography: on action- orientated research. *Progress in Human Geography*, 27(5), pp. 649–657.

Park, R.E., 1928. Human migration and the marginal man. *The American Journal of Sociology*, 33(6), pp. 881–893.

Park, R.E., 1936. Human ecology. *The American Journal of Sociology*, 42(1), pp. 1–15.

Parr, H., 2000. Interpreting the 'hidden social geographies' of mental health: ethnographies of inclusion and exclusion in semi-institutional places. *Health & Place*, 6(3), pp. 225–237.

Philo, C., 2005. Spacing lives and lively spaces: Partial remarks on Sarah Whatmore's Hybrid Geographies. *Antipode*, 37(4), pp. 824–833.

Pieterse, E., 2013. Grasping the unknowable: coming to grips with African urbanisms. In: E. Pieterse and A. Simone, eds., *Rogue urbanism. Emergent African cities*. Auckland Park, South Africa: Jacana Media, pp. 19–35.

Pile, S. and Thrift, N. eds., 1995. *Mapping the subject. Geographies of cultural transformation*. London: Routledge.

Radcliffe, S.A., 2014. Gendered frontiers of land control: indigenous territory, women and contests over land in Ecuador. *Gender, Place & Culture*, 21(7), pp. 854–871.

Rankin, K.N., 2011. Assemblage and the politics of thick description. *City*, 15(5), pp. 563–569.

Rose, G., 1997. Situating knowledges: positionality, reflexivities and other tactics. *Progress in Human Geography*, 21, pp. 305–320.

Roy, A., 2009. Civic governmentality: the politics of inclusion in Beirut and Mumbai. *Antipode*, 41(1), pp. 159–179.

Saldanha, A., 2006. Reontologising race: the machinic geography of phenotype. *Environment and Planning D: Society and Space*, 24(1), pp. 9–24. Available at: <http://www.envplan.com/abstract.cgi?id=d61j> [Accessed 18 Jul. 2012].

Sassen, S., 2014. *Expulsions. Brutality and complexity in the global economy*. Cambridge, MA: Belknap–Harvard.

Saunders, D., 2012. *Arrival city: how the largest migration in history is reshaping our world*. New York: Vintage Books.

Sharp, J., 2009. Geography and gender: what belongs to feminist geography? Emotion, power and change. *Progress in Human Geography*, 33(1), pp. 74–80.

Sidaway, J.D., 2000. Postcolonial geographies: an exploratory essay. *Progress in Human Geography*, 24(4), pp. 591–612.

Simone, A., 2010. *City life from Jakarta to Dakar: movements at the crossroads*. London: Routledge.

Simone, A., 2011. Deals with imaginaries and perspectives: reworking urban economies in Kinshasa. *Social Dynamics*, 37(1), pp. 111–124.

Smith, N., 1998. Giuliani time: the revanchist 1990s. *Social Text*, 57, pp. 1–20.

Smith, R.G., 2003. World city topologies. *Progress in Human Geography*, 27(5), pp. 561–582.

Spivak, G.C., 1988. Can the subaltern speak? In: C. Nelson and L. Grossberg, eds., *Marxism and the interpretation of culture*. London: Macmillan, pp. 271–313.

Spivak, G.C., 1990. Explanation and culture: marginalia. In: R. Ferguson, M. Gever, T.T. Minh-ha and C. West, eds., *Out there: marginalization and contemporary cultures*. New York: The New Museum of Contemporary Art, pp. 377–394.

Staeheli, L.A and Lawson, V.A, 1995. Feminism, praxis, and human-geography. *Geographical Analysis*, 27(4), pp. 321–338.

Stengers, I., 2005. The cosmopolitical proposal. In: B. Latour and P. Weibel, eds., *Making things public: atmospheres of democracy*. Cambridge, MA: MIT Press, pp. 994–1003.

Stonequist, E.V., 1935. The problem of the marginal man. *American Journal of Sociology*, 41(1), pp. 1–12.

Swanton, D., 2010. Flesh, metal, road: tracing the machinic geographies of race. *Environment and Planning D: Society and Space*, 28, pp. 447–466.

Thrift, N., 1996. *Spatial formation*. London: Sage.

Thrift, N., 2005. But malice aforethought: cities and the natural history of hatred. *Transaction of the Institute of British Geographers*, 30(2), pp. 133–150.

Thrift, N., 2008. *Non-representational theory. Space, politics, affect*. London: Routledge.

Thrift, N., 2014. The 'sentient' city and what it may portend. *Big Data & Society*, 1(1), pp. 1–21.

Tissot, S., 2007. The role of race and class in urban marginality. *City*, 11(3), pp. 364–369.

Vasudevan, A., 2014. The autonomous city: towards a critical geography of occupation. *Progress in Human Geography*, pp. 1–22. Available at: <http://phg.sagepub.com/cgi/doi/10.1177/0309132514531470> [Accessed 22 Dec. 2014].

Venkatesh, S.A., 2008. *Off the books. The underground economy of the urban poor*. Cambridge, MA: Harvard University Press.

Wacquant, L., 1996. Red belt, black belt: racial division, class inequality and the state in the French urban periphery and the American ghetto. In: E. Mingione, ed, *Urban Poverty and the Underclass: A Reader*. Oxford: Blackwell.

Wacquant, L., 1999. Urban marginality in the coming millennium. *Urban Studies*, 36(10), pp. 1639–1647.

Wacquant, L., 2002. Review: scrutinizing the street: poverty, morality, and the pitfalls of urban ethnography. *American Journal of Sociology*, 107(6), pp. 1468–1532.

Wacquant, L., 2008. *Urban outcast*. London: Polity.

Weinstein, L., 2014. *The durable slum*. Minneapolis: University of Minnesota Press.

Whatmore, S., 2002. *Hybrid geographies. Natures, cultures, spaces*. London: Sage.

Wylie, J., 2010. Non-representational subjects? In: B. Anderson and P. Harrison, eds., *Taking-place: non representational theories and geography*. Farnham: Ashgate, pp. 99–115.

Recontextualization

2 Grand visions fizzle on the margins of the city

Kavita Ramakrishnan

Introduction

In one of my many conversations with Raiz Khan we fall into a heated discussion about capitalism, as he is curious to understand how I, having grown up in the West, understand it as an economic philosophy and social reality. A Muslim who migrated to Delhi in 1986 from the State of Bihar, and subsequently evicted from his *basti*, or informal settlement, behind the Rajghat Powerhouse, Raiz Khan has survived and even thrived in the resettlement colony compared to his neighbours – in part due to his political savvy, and also due a successful clothing business he started off the colony's main square. He embraces capitalism but he makes clear limits in his defence of it, especially when the conversation veers to the 'world-class city':

> This world-class city thing is only for capitalists. It won't benefit the poor in any way, be it in education or with job opportunities. . . . The government paid no attention and just dumped us here straight away. . . .

I unpack his quote in full in the sections that follow, but use it to frame the emotions that the 'world-class city' – a concept used in Delhi technocrat and design discourse, though a sense of which is largely understood by those I interviewed – elicits amongst the urban poor in relation to shattered aspirations. For Ghertner (2011), the 'world-class city' is a planning and juridical aesthetic discourse pervasive in Delhi that imagines the city as clean, modern, and without slums, a discourse that became widespread in the 2000s. Its coherence on the ground is striking, and though ordinary people refer it to in various ways, it serves as a metonym for exclusionary projects – which evoke ambiguity, despair, pride, and sadness in equal measure amongst my respondents.

I connect Raiz Khan's dismissal of the 'world-class city' to the first part of this chapter's title – 'grand visions fizzle' – taken from a recent photojournalism spread in *The New York Times* (Romero and Berehulak, 2014). The spread focuses on the unforeseen consequences of Brazil's investment in costly and lavish building projects during its economic boom and in the lead-up to the 2014 World Cup. These grand ambitions and the subsequent 'fizzle' are emblematic of the types of speculative urbanism, 'world-class city' projects, and mega-events planning occurring in cities across the world, including Beijing, Cape Town, Delhi, and Rio de Janeiro. 'Fizzle' is a descriptive that captures the deflating aftermath once dreams have launched, and then failed to symbolize or fuel these cities' 'world-class' rise. However, I shift the focus from the 'grand visions' of elite urban actors to the 'grand visions' of the urban poor, with a particular focus on how these change or 'fizzle' in times of insecurity. I am interested in how the 'world-class city' is linked to various frames of progress articulated by the urban poor, what aspects they embrace and/or contest, and how they reconcile exclusionary urban aesthetics with getting ahead.

I draw on ethnographic accounts from a peripheral resettlement colony called Bawana, where people have been forcibly evicted from centrally located *bastis* in Delhi and relocated by the state during the lead-up to the 2010 Commonwealth Games. Based on conversations, I describe the connections and disjunctures residents make between the city left-behind and the resettlement colony. The 'urban', though ambiguous and amorphous – and for those displaced mainly talked about as the distinct perimeter of Old Delhi that circumscribed their previous homes and the central markets – is defined by the ability to improvise and 'make it'. This contrasts starkly with what the 'rural' provides in the minds of respondents. I explore how people 'become' marginal through what they perceive as state neglect and the inaccessibility of Delhi to the poor. The 'city' remains the key point of reference for my respondents as to what constitutes 'ideal' living. Through subjectivities of the 'city left-behind', it becomes apparent that residents are willing to forego certain aspects of their former city life in order to tenaciously grasp the last remnants of urban dwelling. I argue that documenting what the city purportedly offers and means to resettlement residents, even in its more anti-poor manifestations, sheds light on how people conceive of both marginality and belonging. This chapter thus seeks to extend a particular understanding of how different urban marginalities are envisioned in relation to the city, and in relation to urban aspirations.

Aspirations and the city

Appadurai (2013) identifies the textured emotions that accompany the future as a 'cultural form' and describes the wide-ranging sensations of disorientation, awe, vertigo, and excitement that emerge from the many future possibilities. These in turn specifically shape forms of imagination, aspiration, and anticipation. I explore recent work on future making, with the proposition that resettlement and the waiting that ensues represent a particular disjuncture that can drive people to conceive of and act in regards to their future differently than before – and thus differently from the urban poor still precariously inhabiting informal settlements in more central areas of the city. I ground notions of the 'good life' at a more localized and modest level, seeking to understand the personal, familial, and community-level negotiations for different and more desirable futures even amongst the most peripheral – a bottom-up approach to exploring dreams, aspirations, plans, and wants. This discussion yields the following question: what are the futures people envision if they are left out of wider discourses and frames of progress?

Believing that the study of visions of the 'good life' has been relegated to moments of trauma or exception, Appadurai encourages a study of 'a fuller picture in which cultural systems also shape specific images of the good life as a map of the journey from here to there and from now to then, as part of the ethics of everyday life' (2013, p. 292). As has been observed in various contexts, some people cling to exclusionary projects despite the inevitable consequences of dispossession and displacement; simultaneously, reductionist dichotomies are often deployed as 'truths' in talk of the future, through conversations of what can, should, and will happen.

My informants, for instance, would repeatedly describe their lives through highly reductionist categories – centre and periphery, village and city, jungle and civilization, and so on – to explain the differences between the resettlement and informal settlements of the past (Ramakrishnan, 2014). An example of this can be seen from conversations with Feroza, a Bihari Muslim evicted from the Red Fort area of Delhi, who would disdainfully refer to the resettlement colony as a 'village' (*gaon*), but distinguish it from her former village in Bihar – for her, the city (including the peri-urban extensions of the resettlement) allowed her to experience a measure of empowerment and distance from supposedly 'backwards' ideas seen in her natal village. Her views underscore the contradiction that emerges in many of the conversations: Bawana *is* a village, but it is still considered better than respondents' native villages, and in some regards resembles the city (*shehr* in Hindi) more closely. Stating that Bawana is a 'village' is not meant to be a descriptive, but rather a measure of disdain and a modernist teleology of geography, development, society, and individual improvement.

Thus, although much of critical theory focuses on the 'post-modern' and doing away with categories belonging to the ill-conceived legacy of 'modernity', such categories still resonate with people living in the resettlement colony and amongst other groups of 'everyday' people (Ferguson, 1999). 'Development' – as both a material form and abstract plan – has also been documented as positive and a symbol for a better future among ordinary people. Harms (2011), for instance, found that a major road infrastructure project running through Hóc Môn, on the outskirts of Saigon, was associated with the larger, desired project of 'development' despite the negative impact it had on the local community. People he interviewed would compartmentalize the frequent deaths and injuries related to the road construction, lauding instead the promise the completed project held – in effect suffering through development was seen as necessary to achieving the coveted end goal of improved living conditions. Writing on urbanization in Kinshasa, De Boeck (2011) comes to similar conclusions: the implementation of 'world-class' developments speaks to what people are willing to lose on the gamble of becoming a 'part' of the city. The grandiose schemes for the 'new' Kinshasa include an exclusive development of private villas on an artificially created island on the Congo River. Rather than expressing dissatisfaction or anger with the exclusionary nature of this project and the potential for displacement, the farmers whose livelihoods come from agriculture on the riverbanks have embraced these speculative visions. De Boeck explains the resonance that exclusionary ideals hold for everyday residents of Kinshasa as an awakening, ironically, of the 'capacity to believe' in a different urban life. This dream is one that is held across socio-economic classes – the promise of beauty and betterment in the city is so alluring that it trumps the potential struggle and devastation it leaves in its wake.

In interviews with farmers who faced the threat of relocation if the plans came to fruition, De Boeck records their responses to his query of whether they understood the full implications of the project as the following: 'Yes, we'll be victims, but still it will be beautiful' (p. 278). This striking statement signifies the partial willingness of the urban poor and working classes to bear the brunt of speculative futurity and highlights the contradictions in this particular form of belonging. De Boeck sees the projects most likely stalling in the planning phase, but he locates the power of the words, images, and visions as the 'ultimate weapon' through which people confront the harshness and hostility that can be met within the city.

Similarly, in the context of Delhi, Baviskar (2010) talks about the power certain visions of urban regeneration related to the 2010 Commonwealth Games preparation held for ordinary residents. Even though infrastructure investment and promises of job creation related to the 'extraordinary' event had little impact – and oftentimes negatively affected the urban working classes – the compelling vision for the city initiated a general 'urban buzz' over the planned changes that all could share (Baviskar, 2010, p. 158). She claims the Games introduced a sense of excitement over the future of the city as it bound citizens together across different social strata. Ghertner (2011) observed a similar phenomenon when interviewing residents of informal settlements displaced in preparation for the Games. Although the residents expressed their devastation at being evicted, many of them also vocalized the necessity for the clearance of their homes in order for the Games to successfully proceed. In one account, a woman calmly and without irony articulates her vision of a future Delhi: 'Delhi will be a beautiful city, totally neat and clean. All the slums will be removed and there will only be rich people' (Ghertner, 2011, p. 15). Thus, aspirations for an ordered Delhi exist even among inhabitants of informal settlements, exhibiting what Ghertner terms as the popular cultivation of a 'world-class aesthetic' in spite of the consequences that such an ordering has for their dwelling and livelihoods. The adoption and internalization of 'world-class aesthetics' amongst slum dwellers becomes a 'basis for both locating themselves in the changing city and for framing their own world-class aspirations' (p. 281). In this sense, the urban poor for Ghertner do not just passively watch developments take place around them, but, through their desires, express particular forms of agency – in particular, a motivation to move beyond life in the *bastis* or even the desire for the elimination of such dwelling forms, as they feel they have no aesthetic place in the city. Baviskar (2010) and Ghertner's (2011) work thus suggests that the internalizing of 'world-class aesthetics' and developments among citizens of Delhi signifies the permeation of dominant future city ideals (cf. Jervis-Read, 2010; Datta, 2012).

It is at this juncture I position my research by exploring how people pushed to the margins of the city situate themselves within dominant futures and what their own future imaginaries entail. However, while Harms (2011), De Boeck

(2011), Baviskar (2010), and Ghertner (2011) have captured a particular aspect of development, namely the aspiration for beautification in city spaces amongst the poor, I move away from their claims that the poor maintain a sense of calm and inevitability at their impending victimhood. While better roads and parks are welcomed amongst my respondents, there is also disbelief and anger that such developments happen at the tremendous cost of their own livelihoods, homes, and lives. Interestingly, I observed not one or the other response but the (almost paradoxical) intertwining of the two: often a simultaneous praise for the city's beautification efforts and dismay that beautification has become equated with slum clearance. The city proper – even in its 'slum' form – though precarious, is a distinctive space that offers a realm of possibilities for my respondents, who migrated to Delhi precisely in search of such possibilities. To be removed from such a space represents a severe, and, in many cases, unrecoverable material and psychological blow to people who had invested their savings and rooted their aspirations in the city.

I use Simone's (2010) 'politics of anticipation' here as a way of understanding what the city means to the urban poor and challenging the importance that the existing literature cited above places on the embrace of urban aesthetics by the poor. The 'politics of anticipation' refers to a specific temporality that is heavily underlined by risk but open and flexible to the multiplicity of opportunities present, encompassing strategies of entrepreneurship and survival that focus on action in the moment with the resources available. It is a politics that, according to Simone, emerges alongside new forms of urban experimentations and calculations and is tied up with the distinctive space of the city with its plurality of opportunities for networking, brokering, and leveraging. The provisionality of the city described by Simone (2004) lends itself to the shrewd improvisation, termed variously as *jugaad*[1] and 'hustle',[2] that is the staple of the 'underground' economies present in these inner-city and low-income neighbourhoods. *Jugaad* thrives in the urban context because of the realm of possibility that allows one to act in new and improvisational ways. At the same time, Simone underscores that regularity is a crucial ingredient to making provisionality productive: familiar experiences and perceptions strengthen the desire of individuals to circulate amongst different actors and neighbourhoods, and, over time, they reference these interactions and transactions. As he notes, '[Even] when actors do different things with one another in different places, each carries traces of past collaboration and an implicit willingness to interact with one another in ways that draw on multiple social positions' (Simone, 2004, p. 408). What presents itself in the indefinable space of the city is the opportunity to experiment, and, despite the often limited avenues for expanding one's trade against competition, the 'politics of anticipation' (Simone, 2010; Simone and Fauzan, 2013) remains viable in the city – more specifically, there is room for 'canvassing the actual initiatives of others . . . sensing what could have been done, what could have taken place' (Simone and Fauzan, 2013, p. 121). This enables residents, according to Simone and Fauzan (2013), to attempt different initiatives without the pressure of committing to or immediately relying on employment from more established ventures. I link this to *jugaad*/hustle, in that the improvisation can occur through various channels, avenues, and actors, without the reliance on established forms of working or doing.

'Delhi is beautiful, but what good is this for us?': 'world-class' subjectivities and city imaginaries

The following profiles draw from specific questions on how people perceived the city and their place in it, particularly after experiencing brutal eviction and resettlement. At the beginning of the Games preparations in 2004, evictions of more than 27,000 families (close to 90,000 people) from the Yamuna Pushta (River Yamuna banks) area occurred in February and April as residents were accused of unduly polluting the Yamuna River, amongst other allegations (Menon-Sen, 2006; Padhi, 2007; Srivastava, 2009). Many were shocked by the eviction, as they believed that the courts would ultimately stay the demolitions given the communities' existence on that land for more than three decades in Delhi (Menon-Sen, 2006).

Resettlement colonies founded in the past 15 years are located 35 kilometres on average from the former homes of *basti* residents. The cost and duration (often up to three hours) of commuting this distance means that previous sources of employment are no longer tenable. Resettlement was also not guaranteed to all *basti* dwellers: families were required to demonstrate eligibility through ration or voter identity cards, proving residence in Delhi either before 1990 or 1998, and pay a plot lease fee of 7,000 rupees.[3] As a result, more than 50% of those evicted were rendered homeless (Menon-Sen, 2006; Dupont, 2008). Those who managed to secure housing lament the government's failure to provide proper roads, drains, or consistent access to water in the resettlement colony. Even after ten years, many residents comment that life in the resettlement cannot compare to the location, infrastructure, amenities, and labour opportunities to which life in *bastis* provided access.

'They should have made something for us . . . '

Sandeep is a 38-year-old Hindu woman, born and raised in Delhi, and evicted from Yamuna Bazaar (YB). Extremely generous, Sandeep always sends her son to buy us a litre of soft drink during our chats, pulling wads of money from inside her blouse. Like many of the widows I interviewed in the colony, she supplemented her income by renting out the ground floor of her house to a shopkeeper. In conversations, Sandeep often wistfully talked about Delhi as a city that was beautiful from the beginning of her life there, an 'amazing' city from which Bawana is completely separate. Nobody could starve there, she felt, as the potential existed to earn money and feed one's kids. As Simone (2004, p. 407) notes, the marginality of the inner city is often countered by the diverse array of people, or the 'intersection of socialities' that make for collaboration and possibility. Sandeep in particular found value in what she claimed to be a highly mobile lifestyle within and around the *basti* and the social and economic connections she was able to make in the larger urban sphere.

Even though she has a comparatively well-built house in Bawana, she reiterated that, if given a chance, she was ready to move back for even six yards of space (which is half of what she currently owns). In her next statement, however, she quickly rejected the possibility of ever achieving such a dream: to her, Delhi had become a place where only rich people, possessing at least one or two crore rupees (or the equivalent of 200–400,000 US dollars) could live well. In her mind, the government could not be considered pro-poor, as she and others had been living peacefully in YB and were still evicted. She scornfully responded to a question about her perceptions of the Delhi government by saying, 'What is there left to think of a government that ruins poor people completely?' It did not make sense to Sandeep that Delhi is considered a 'world-class city', especially if the government is targeting poor people for displacement through its beautification processes. Sandeep believed the government could have instead approached eviction and resettlement through a more middle-ground approach: by taking people scattered in YB and building one residence for them, officials would thus free up the rest of the land for governmental use:

> By kicking and making the poor folks suffer, if the government intends to make something, it is not constructive. You kick poor folks [out] and then wish to construct? If they intended to make [a development], then somewhere in the corner they should have made something [for us] – a flat maybe, where we poor people could be given some place. [They could have] made us all shift there [itself], somewhere in the corner . . . the public would be happy as well.

Interestingly, Sandeep alluded to the lobbying of various groups (whom she terms 'the public') to remove the settlements along the river, and she believed that the government unfairly capitulated to the demands of others without providing a suitable alternative to the residents. Sandeep felt further misled by the government since despite its post-eviction plans of developing the land she had been residing on, she had seen this land lying vacant only recently.[4] In addition to what she sees as a locational downgrading, Sandeep pointed to the considerable size difference in her housing and infrastructural deficiencies: the government allocated 12.5 square yards in Bawana, whereas in YB, Sandeep claimed, her kitchen alone was that size; similarly, she remembered the infrastructural connections provided by the government as far better than those in Bawana, with water, electricity, common taps, and hand pumps present and working in YB. Although the former may be an exaggeration, particularly of the space to which she previously had access, Sandeep conveys a deep distaste of the government's 'solution' for the poor.

Sandeep's opinions on Delhi convey an inherent contradiction: despite her perceptions of the city's treatment of the poor, she would prefer returning there, even if it meant living in a considerably smaller dwelling. While acknowledging that Delhi is a city for the wealthy, she still expressed ambitions, however lofty, to be a part of this exclusionary urban space. Although she is not dreaming against all odds of a lifestyle similar to urban elites, she longs for the choice of residing, even in an impoverished state, amidst the spectacle of Delhi futurity. I differentiate my findings from the works cited earlier by emphasizing an important nuance: Sandeep cannot reconcile that she and many others had to bear the brunt of this urban futurity through the violence enacted upon their bodies and livelihoods.

'Birds in a cage'

Mohib Kalam, a man in his mid-fifties who was displaced from Yamuna Pushta, spoke scornfully of the treatment of the poor in the new and improved Delhi. For Mohib Kalam, Delhi became beautiful, but the government ruined his life and that of others in the process. Migrating to Delhi from North Bengal in 1980, he found the situation for migrant labour relatively stable and recalled how factory owners would welcome labourers, with factories mushrooming in areas such as Kashmere Gate and the New Delhi Railway Station. In contrast to the situation he sees now, Mohib Kalam remembered labourers as being valued for contributing to the daily functioning of the city, often provided with room and board by employers. Although he changed jobs regularly (from owning a jewellery/handicrafts stall, to a stationery shop, to a tea stall) and faced his own set of difficulties, he said he never felt the same precariousness he faces currently in the resettlement colony. According to him, Delhi was a city where if a person possessed cunning (*chaal*) to get by and get work done, strength (*baal*) to challenge opponents, and intelligence (*akal*), s/he could experience a certain measure of success. To some extent, Mohib Kalam treats 'playing the game' in Delhi with nostalgia: a game not without risks and often straddling the boundaries of legality and illegality, but accessible even to new migrants. Bawana is unsettling, as it does not afford the same spontaneous opportunities to make money that he found in Delhi; instead, he is one of many who finds the above qualities of little use in the face of temporal ruptures. I return to the earlier discussion of *jugaad* (shrewd improvisation) and how the provisionality of the city leads to particular forms of *jugaad* (Simone, 2004, 2010) – the city as a site of improvisation is not lost on Mohib Kalam and others who struggle to find the same range of channels and actors through which to relaunch their ambitions in the resettlement colony.

Mohib Kalam sadly stated that, as a young man he had grandiose dreams of wealth and success in Delhi, and this dream had sustained him while living in Yamuna Pushta; now, he realizes that he failed to accomplish anything substantial. He complained about his current occupation as a mini-bus driver for residents commuting to Delhi, leaving in the early hours and bringing them back late at night. In the face of what he sees as a dead-end job, Mohib Kalam planned to instead return to his native village, as he believed his prospects there to be marginally better. He was one of the few respondents I met who actively longed to go back to North Bengal, demonstrating that not all residents in the colony experience a permanent relationship with the city, and that other sentiments about the 'village' exist beyond the main tropes discussed earlier. However, although he considered his village a 'place of the heart', by his own admission, had he remained in Pushta, Mohib Kalam might have considered permanently staying in the city.

Like Sandeep, he believed that the state could have resettled people in Yamuna Pushta itself, as plenty of space existed, according to him, even with the planned building of the park and flyover. He rhetorically demanded to know the logic behind building another park when so many existed, including Rajghat, Vijayghat, and Shastri Park. He answered his own query with the derisive statement that the government wanted to 'fake it before the foreigners' – that the dirty poor did not exist in Delhi. He told me he could no longer enjoy beautiful public spaces in Delhi, as he felt that the government would be ready to evict him from those too. However, unlike Sandeep, the beautification of Delhi was worthless to him, since he could not enjoy it with his children. Yes, he said, Delhi became beautiful, but the government ruined their lives in the process, creating a living hell for them in Bawana. Delhi, he mournfully noted, is only beautiful for rich people. Mohib Kalam said that he used to live near major shopping areas and parks, and it was easy to take his children for a picnic; now he cannot think of this because of the overcrowded and lengthy bus journey to Delhi filled with pickpockets. He further described himself as part of the lowest class, or in his words a 'bird in a cage' that could not access 'world-class' developments due to a lack of money:

> If you see through the eyes of the poor, you will find that they feel like birds in a cage with lots of luxuries such as fruits and gardens on the outside. But the fact is they will get nothing from there – they cannot enjoy it. We can only crave it. We can only see the skyscraper, but cannot enter it. We don't even have a home or proper clothes. If this [Bawana] becomes the world-class city then we would be told to leave this place as well.

His cynical prediction that if beautification happened in Bawana, he and others would find themselves in the same situation is telling: Mohib Kalam no longer has any faith in the government's treatment of the poor. He concluded that the country and Delhi were developing, but not Bawana. From his quote, Mohib Kalam also maintained that the poor 'crave' the same amenities as those afforded to the rich, a point worth emphasizing: the idea of elite-style beautification and luxury is not criticized here and is instead seen as aspirational. Indeed, the skyscraper has, in many cities, been created and seen as the ultimate symbol of neoliberal global modernity, and as a vertical icon of power and futurity (see Graham and Hewitt, 2013). Its reference in this conversation signals a particular imaginary of an exclusionary space that is starkly juxtaposed with an informal settlement. Mohib Kalam's use of the skyscraper as the metaphor of choice speaks to a certain purchase of such exclusionary forms, even in face of the 'perpendicular splintering' (Graham and Hewitt, 2013) they cause.

'This world-class city thing is only for capitalists'

In relation to the 'world-class city', others blame 'capitalists' for turning out the poor from Delhi and ensuring that the labouring classes can no longer make a home in the city. Raiz Khan, a Bihari Muslim who migrated to Delhi in 1986, and built a *jhuggi* in Indira Colony, close to the Rajghat Powerhouse Station, feels as though he should leave Bawana but his relative success compared to his neighbours is still not enough to go elsewhere. Despite being quick to highlight the endless struggles of living in his previous *basti* (for instance, he described the constant flooding that the *jhuggis* faced), the ongoing corruption and crime in Bawana surpassed anything he had experienced in the past – and more importantly, posed a considerable threat to the future of his children. His frequent trips to the markets of Chandni Chowk in Old Delhi to procure clothing for his dress shop in the colony meant that he was in constant contact with what he terms a 'better and safer environment', one that he wished he could move back to with his seven children; however, Raiz Khan was convinced that such a desire would remain an unfulfilled dream.

He would frequently bring up what he saw as the overtly capitalist nature of the Delhi economy, which in his definition disregarded the essential labour that many residents had formerly provided for middle-class citizens, thus destroying the true 'ethos' of capitalism (defined by him as the circulation of labour opportunities). Raiz Khan gave the example of the location of resettlement colonies, which he felt meant that capitalist drivers (both politicians and industrialists) of the city were colluding to move the poor to places where their labour could be exploited. In this case, he was specifically referring to the proximity of the Bawana Industrial Area to the resettlement colony, although Raiz Khan also cited other examples such as Sultanpuri, Narela, and Wazirpur.[5] In his own words:

> This world-class city thing is only for capitalists. It won't benefit the poor in any way, be it with education or with job opportunities. . . . The government paid no attention and just dumped us here straight away. The rulers of any country have to take care of the poor as well since they are needed equally. Schools and offices need peons, and houses need maids. But this idea of development is absent in the government. They use the poor but only to ensure that their standards are not raised enough to reach their status. This is the ladder of capitalism.

Although Raiz Khan maintained that capitalism is a better system than socialism (he discussed in detail the failures of the Russian 'experiment'), he felt that in its current form it only succeeded in turning millionaires into billionaires without any developmental efforts to emancipate the living standards of the poor. He seemed to believe, however, that the labour stratifications he identified, such as the office peons and housemaids, were acceptable to the workings of a particular form of capitalist economy – a form seen in the Delhi that still acknowledged the value of working-class labour pre-evictions. It is the 'new' form of capitalism in the 'world-class city that he condemns, where inequalities lead to the erasure of this labouring class from the city limits. The 'world-class city', in Raiz Khan's mind, was guilty then of taking capitalism a step too far, by crushing the living and working spaces of the poor.

Conclusion

Although respondents' criticism, such as those by Sandeep and Mohib Kalam, of the 'world-class city' and its related projects capture the unjust actions of the state during and after eviction, they reproduce particular discourses of urban beautification. Many of my interviewees do not offer a counter-hegemonic response to the growing exclusivity of Delhi, instead largely focusing on how they no longer can even reside in the newly beautified city. The city, then, remains an aspirational home for the respondents despite their admissions that they would little enjoy or be able to access such exclusionary spaces – living in Delhi is still something these respondents craved, no matter how small or diminished their dwelling became. Others, such as Raiz Khan, although condemnatory of the 'world-class' city's impact on the poor, are assenting of larger ideological projects (such as the labour stratification of the working classes), which often have exploitative outcomes for the impoverished. The urban poor thus cite multiple reasons for their dispossession that reveal particular understandings of urbanity, development, the state, and social justice – and not necessarily in progressive or radical ways (cf. Baviskar, 2010; De Boeck, 2011; Ghertner, 2011; Harms, 2011).

However, the assumption that the urban poor embrace city beautification is an incomplete and inaccurate picture, as they know full well the attendant sacrifices. The story is instead more nuanced, and my research demonstrates that the city is of irreplaceable value – Delhi's beautification means nothing if they are not a part of it. Removed and excluded from the developments taking place within the city, respondents still frequently praised the transformations in Delhi and referred to the city as 'beautiful', among other descriptions. Yet an admiration of Delhi's green spaces, expressways, and new malls turned into a bitter reflection of how the poor are no longer wanted within the confines of such a city. Returning to Ghertner's (2011) quote, the acceptance of 'world-class aesthetics' by *basti* residents apparently forms the 'basis for both locating themselves in the changing city and for framing their own world-class aspirations' (p. 281). There is another way of understanding this – through the lens of Simone (2004, 2010) and Simone and Fauzan (2013) – where, instead of 'world-class aesthetics' shaping and defining aspirations of the urban poor, the ability to 'make it' and survive and potentially thrive play a much larger role. For Sandeep and Mohib Kalam the connections and opportunity to innovate in the city counteract destitution in ways that the resettlement colony cannot afford. Similarly, Raiz Khan criticizes the difference between the capitalism he feels brings benefit to the poor, in terms of employment, compared to the purely exploitative form he sees through the 'world-class city'. This examination of varying shades of exploitation and minimal claims on urban space is also crucial to understanding marginality; namely, how people who are resettled frame their place in an exclusionary city. It is not the remnants of beautification that the urban poor claim during their inner-city dwelling, but the possibility of social and economic transformation – something they mourn the loss of in the resettlement colony.

Acknowledgements

Many thanks to Michele Lancione and Eszter Kovács for their thoughtful comments. I also am indebted to Pushpa Arabindoo for pushing me to think about aspirations in more radical ways. I save my final thanks for the residents of Bawana who continue to humble and inspire me.

Notes

1 *Jugaad* is a colloquial Hindi word that refers to the Indian notion of improvisation, or 'an innovative fix; an improvised solution born from ingenuity and cleverness' (Radjou et al., 2012, pp. 4–5). However, researchers see *jugaad* exhibited by people in a variety of ways, ranging from forms of shrewd improvisation and quick thinking to practices such as bribery and corruption (see Jeffrey, 2010; Jeffrey and Young, 2014; Jauregui, 2014).
2 The word 'hustle', with its origins in African-American culture, is increasingly used in ethnographic contexts to describe new and innovative forms of economic activities (see Thieme, 2010 and Alpes, 2014).
3 This is close to a month's wage for a colony resident and, as of 4 July 2015, is equivalent to 110 USD/71 GBP.
4 Both Dupont (2008) and Bhan and Shivanand (2013) have found a high percentage (close to 50% in the case of Dupont's sampled sites) of sites formerly occupied by squatters still lying vacant post-eviction.
5 Resettlement colonies in Sultanpuri, Narela, and Wazirpur (created at different times: during the Emergency and during the past ten years) are located close to industrial sites.

References

Alpes, M.J., 2014. Imagining a future in 'Bush': migration aspirations at times of crisis in Anglophone Cameroon. *Identities: Global Studies in Culture and Power,* 21(3), pp. 259–274.

Appadurai, A., 2013. *Future as cultural fact: essays on the global condition.* London and Brooklyn: Verso.

Baviskar, A., 2010. Spectacular events, city spaces, and citizenship: the Commonwealth Games in Delhi. In: J.S. Anjaria and C. McFarlane, eds., *Urban navigations: politics, space and the city in South Asia.* London: Routledge, pp. 138–164.

Bhan, G. and Shivanand, S., 2013. (Un)settling the city: analysing displacement in Delhi from 1990 to 2007. *Economic & Political Weekly,* 48(13), pp. 54–61.

Datta, A., 2012. *The illegal city: space, law and gender in a Delhi squatter settlement.* Farnham: Ashgate Publishing Limited.

De Boeck, F., 2011. Inhabiting ocular ground: Kinshasa's future in light of Congo's spectral policies. *Cultural Anthropology,* 26(2), pp. 263–286.

Dupont, V., 2008. Slum demolitions in Delhi since the 1990s: an appraisal. *Economic & Political Weekly,* 43(28), pp. 79–87.

Ferguson, J., 1999. *Expectations of modernity: myths and meanings of urban life on the Zambian Copperbelt.* Berkeley and Los Angeles: University of California Press.

Ghertner, D.A., 2011. Rule by aesthetics: world-class city making in Delhi. In: A. Roy and A. Ong, eds., *Worlding cities: Asian experiments and the art of being global.* Oxford: Wiley-Blackwell, pp. 279–306.

Graham, S. and Hewitt, L., 2013. Getting off the ground: on the politics of urban verticality. *Progress in Human Geography,* 37(1), pp. 72–92.

Harms, E., 2011. *Saigon's edge: on the margins of Ho Chi Minh City*. Minneapolis: University of Minnesota Press.

Jauregui, B., 2014. Provisional agency in India: *jugaad* and legitimation of corruption. *American Ethnologist,* 41(1), pp. 76–91.

Jeffrey, C., 2010. *Timepass: youth, class, and the politics of waiting in India.* Palo Alto: Stanford University Press.

Jeffrey, C. and Young, S., 2014. Jugād: youth and enterprise in India. *Annals of the Association of American Geographers,* 104(1), pp. 182–195.

Jervis-Read, C., 2010. Making Delhi like Paris: space and the politics of development in an East Delhi resettlement colony. PhD Diss., University of Sussex.

Menon-Sen, K., 2006. 'Better to have died than to live like this': women and evictions in Delhi. *Economic & Political Weekly,* 41(20), pp. 1969–1974.

Padhi, R., 2007. Forced evictions and factory closures: rethinking citizenship and rights of working class women in Delhi. *Indian Journal of Gender Studies,* 14(1), pp. 73–92.

Radjou, N., Prabhu, J. and Ahuja, S., 2012. *Jugaad innovation: a frugal and flexible approach to innovation in the 21st century*. Noida: Random House Publishers.

Ramakrishnan, K., 2014. Disrupted futures: unpacking metaphors of marginalization in eviction and resettlement narratives. *Antipode,* 46(3), pp. 754–772.

Romero, S. and Berehulak, D., 2014. Grand visions fizzle in Brazil. *The New York Times,* 12 Apr.

Simone, A., 2004. People as infrastructure: intersecting fragments in Johannesburg. *Public Culture,* 16(3), pp. 407–429.

Simone, A., 2010. *City life from Jakarta to Dakar: movements at the crossroads*. New York: Routledge.

Simone, A. and Fauzan, A.U., 2013. Majority time: operations in the midst of Jakarta. *The Sociological Review,* 61(S1), pp. 109–123.

Srivastava, S., 2009. Urban spaces, Disney-divinity, and moral middle classes in Delhi. *Economic & Political Weekly,* 44(26–27), pp. 338–357.

Thieme, T., 2010. Youth, waste and work in Mathare: whose business and whose politics? *Environment and Urbanization,* 22(2), pp. 333–352.

3 After a revolution

Public spaces and urban practices in the core of Tunis[1]

Francesca Governa and Matteo Puttilli

Introduction

14 January 2011 is the most representative day of the Tunisian Revolution, which led to the fall of the regime of Ben Ali. The great demonstration on the Avenue Habib Bourguiba, in Tunis, which began early in the morning and lasted until late in the evening, forced the President to flee the country. More or less detailed reconstructions of the preconditions (Mabrouk, 2011) and of the events and protagonists of the Revolution are easy to find on the Web, in newspapers, and in scientific journals (Bettaïeb, 2011; Weslati, 2011), and there are many analyses of the effects of the Revolution at different levels (Gobe, 2012).

The Revolution was a *break* in the political geography of the Mediterranean region. A country regarded as being at the margins of the global geopolitical order has acquired the role of *forerunner* of the so-called Arab Spring that has spread through many other cities of the Maghreb and the Middle East and inspired, at last, different forms of occupation, protest, and revolt in Europe and around the world (Harvey, 2013). However, the Tunisian Revolution, as well as other Arab revolutions, primarily shows how a marginalized population – ruled by a violent, authoritarian, and repressive regime – has been able to protest against a condition of political exclusion and social control, collectively claiming and asserting its right to existence. According to Saskia Sassen (2011, p. 574), this is not simply a matter of (gaining) power: '[M]any of the protest movements we have seen in North Africa and the Middle East [. . .] may not have gained power, they are still powerless, but they are making history and politics'. Their marginal condition (or the *powerlessness*) is therefore not an economic and social disadvantage. Or, at least, it is not only this. It was not only the people belonging to the poorest strata of the population who participated in the Revolution, but, in fact, it was the whole Tunisian society, in its various components and with its social stratification, who rebelled, demonstrated in the streets, and occupied squares.

More than 40 years ago, Henri Lefebvre (1970) wrote that either the revolution will be urban or it will not be. The space, and more precisely the urban space, is both one of the means to transform society and a test of the changes to it, which may not be such if it does not also transform spatial relationships. References to Lefebvre appear in many interpretations of the Arab Spring, highlighting the necessarily urban character of the revolutionary events of 2011 (Lopes de Souza and Lipietz, 2011; Lipietz and Lopes de Souza, 2012), the performances of space (Gregory, 2013), and the *change of life* through the *change of space* in Cairo and Tunis (Kanna, 2012). Urban space is thus considered to be a constitutive element of political actions that are not necessarily urban oriented, but which *use* the city and its spaces to build, strengthen, and assert more general political claims. Assigning a constituent value to the conflict, these interpretations view cities as malleable and contested, and as the very space of politics (Brenner, Marcuse, and Mayer, 2012; Magnusson, 2014) in which public spaces reemerge as symbols of a renewed social and political vitality. According to Swyngedouw (2001, p. 3), '[P]olitics appears as a practice of reorganizing space under the aegis of equality; it emerges where it is not supposed to be, in public space'. In Tunis and in Cairo, 'public spaces in a weak sense turned into public spaces in a strong sense, as they turned politically *vital*; public spaces reinvented as spaces of socio-political contestation, reinvigorating past iterations of protest' (Lopes de Souza and Lipietz, 2011, p. 621).

We arrived in Tunis in April 2013, two years after the Jasmine Revolution, with the aim of exploring the symbolic spaces where the uprising took place. The approach we chose was to immerse ourselves in the public spaces of the central core of Tunis, following the everyday practices in their making and recording and assembling things, speeches, tools and social relations, words and images, sounds and sensations. The result of this exploration is a multimedia bilingual (Italian and French) Web-research entitled *Al centro di Tunisi. Geografie dello spazio pubblico dopo una Rivoluzione – Au centre de Tunis. Géographies de l'espace public après une Révolution.*[2]

In this text, we focus on some issues of this research and, in particular, on the role of space, especially urban public space, as a physical and symbolic place of social and political change. The chapter presents the public spaces of the centre of Tunis where the main events of the Revolution of 2011 (demonstrations, occupations, mass gatherings, etc.) took place, with the aim of investigating the memory of what happened. Following the observation of the practices taking place during our stay in Tunis, we wanted to identify the signs of a change in the perceptions and the uses of public urban space, which from being a space of injustice, inequality, and oppression, is then perceived, conceived, and lived (to quote the Lefebvrian triad; Lefebvre, 1974) as a space of contestation, openness, and possibility. Exploring the practices that occur in the public space two years after the Revolution becomes a way to *bring into existence* not only a political change but also a social and spatial change.

After the introduction, the second section resumes the main spatial-temporal sites of the Tunisian Revolution, emphasizing the role of urban spaces in the events leading to the Revolution's fulfilment. Subsequently, the third section presents the geography of what happened in April 2013 in the three main public spaces in the centre of Tunis (the Avenue Bourguiba, the Place de la Kasbah, the Medina), chosen for their physical and symbolic meaning during the uprisings and, more generally, for their role in the socio-spatial organization of Tunis.

Finally, the fourth section highlights the continuous *rewriting* of the geography of these public spaces through new practices, renewing their timing and modes of attendance, and constantly changing the forms and social, political, and cultural meanings of urbanity.

(Looking for) the urban memory of the Jasmine Revolution

The Tunisian Revolution did not start in Tunis nor on 14 January. On 17 December 2010, in Sidi Bouzid, Mohamed Bouazizi, a young graduate unemployed street vendor of fruit and vegetables set himself on fire outside the headquarters of Governorate to protest after a quarrel with the police. The self-immolation of Bouazizi went around the world and unleashed, in Tunisia, a series of uprisings and protests that within a few days had blended together and given rise to a revolutionary movement.

In the following days, in Sidi Bouzid, and particularly in the marketplace where Bouazizi was working, spontaneous demonstrations and clashes with security forces took place. Then, the uprisings reached other interior cities (Menzel, Bouzaiane, Kairouan, Sfax, Kasserine) and finally arrived in Tunis: sit-ins and demonstrations were held in front of the Palace of Justice, near the Place de la Kasbah, in front of the headquarters of the UGTT (*Union Générale Tunisienne du Travail*) near the Avenue Bourguiba, in the suburbs of Kram, La Fayette, and Ettadhamen. These protests culminated in the general strike of 14 January that brought about 60,000 people onto the Avenue Bourguiba. The demonstration, at first festive and peaceful, was by the afternoon attacked by the police. A few hours after the start of the curfew, set at 5:00p.m., the state television gave the news of the flight of the President from the country.

Despite the end of the regime, Tunis continued to be crossed by riots and demonstrations, which revealed the first divisions related to the management of the Revolution and its future inheritance. The formation of the new government led by Mohamed Ghannouchi – where the majority of the members are of the *Rassemblement Constitutionnel Démocratique* (RCD), the party founded by Ben Ali – triggered a new wave of protests. *Caravanes de la Liberté* (Convoys of Freedom) left from the cities of the interior in which the Revolution had started. They led tens of thousands of people to Tunis, to the Place de la Kasbah. The occupation of the Place, called Kasbah 1, took place peacefully and the *occupied Kasbah* became the symbolic and physical place where the counter-revolutionary power expressed itself. This role was also reaffirmed by a second occupation, called Kasbah 2, in which more than 100,000 people demanded the dissolution of all the institutions of the previous regime, the outlawing of the RCD, and the establishment of a Constituent Assembly. Following the resignation of Ghannouchi, the government was assigned to Beji Caid Essebsi, a politician of the Bourguiba era, while the Constituent Assembly elections, which took place between 20 and 23 October 2011, resulted in Ennadha, a party of religious inspiration, winning.

The continuous mobilization with the contribution of all the Tunisian people started protests and nurtured demonstrations and riots. But it is in Tunis, and especially in its central public spaces, that the Tunisian Revolution found its landing place and its fulfilment. Here, as in other capitals of the Arab Spring, the city, both as a whole and some specific places within it, is not just a passive *scenario* for action. On the contrary, urban spaces are active players of the change, the role of which is expressed in making possible encounters, comparisons, learning, and exchange, among a multiplicity of subjects; in fostering the construction and recognition of symbols as well as ties – both unstable and changeable – that hold together 'a vast diversity of ages, politics, religions, cultures' (Sassen, 2011, p. 575); in allowing the discovery and deployment of practices of conflict and protest; in constituting a 'space of appearance' (Butler, 2011), where the powerless can become present and visible to each other and thus gain aware ness and consciousness of 'the possibility of making the political, or making the civic, or making history' (Sassen, 2011, p. 574). The social movements that made the Revolution in 2011, in Tunis and in other cities of the Arab world, do not have clear leadership and structured hierarchies (Filiu, 2011). They seem to act, echoing Deleuze and Guattari (1972), as a 'body without organs'; that is, as 'an assemblage that gathers people into meaningful relations with each other' and that 'has enough consistency to hold together, to function in some meaningful way as a body politic, as a coherent polis' (Purcell, 2013, p. 124). Urban squares in Tunis, as Tahir Square in Cairo, Pearl Roundabout in Manama, Green Square (today Martyr's Square) in Tripoli, Freedom Square (today Change Square) in Sana'a, and – although not directly involved in the Arab Spring – Taksim Square in Istanbul and Tehran's Revolution Square in Tehran are all examples of democratic assemblages, spatiotemporal places 'to quite literally build an alternative habitus where the act of occupation becomes the basis for producing a common spatial field in which the ongoing interactions of participants continually produce sentiments, ideas, values and practices' (Vasudevan, 2014, p. 3). It is a sort of collective subjectivity that goes beyond the boundaries of the space and time of the revolt and becomes *transposable* to other movements and spaces of protest (Butler, 2011; Gregory, 2013), as well as part of a shared reworking of the Revolution through the change of post-revolutionary urban practices.

Wandering in the core of Tunis, two years after the Revolution

What happened in the public spaces at the centre of Tunis two years after the Revolution? The Avenue Bourguiba, the Place de la Kasbah, and the Medina are the spaces we explored and wandered around during our fieldwork in April 2013 (Figure 3.1). They are very different spaces and were differently invested with meaning by the revolutionary events of 2011.

The street where 'everything happens'

Avenue Bourguiba is spoken of and elected, as a whole, as the street where 'everything happens'. Even if the Revolution was born in the cities of the very inland of

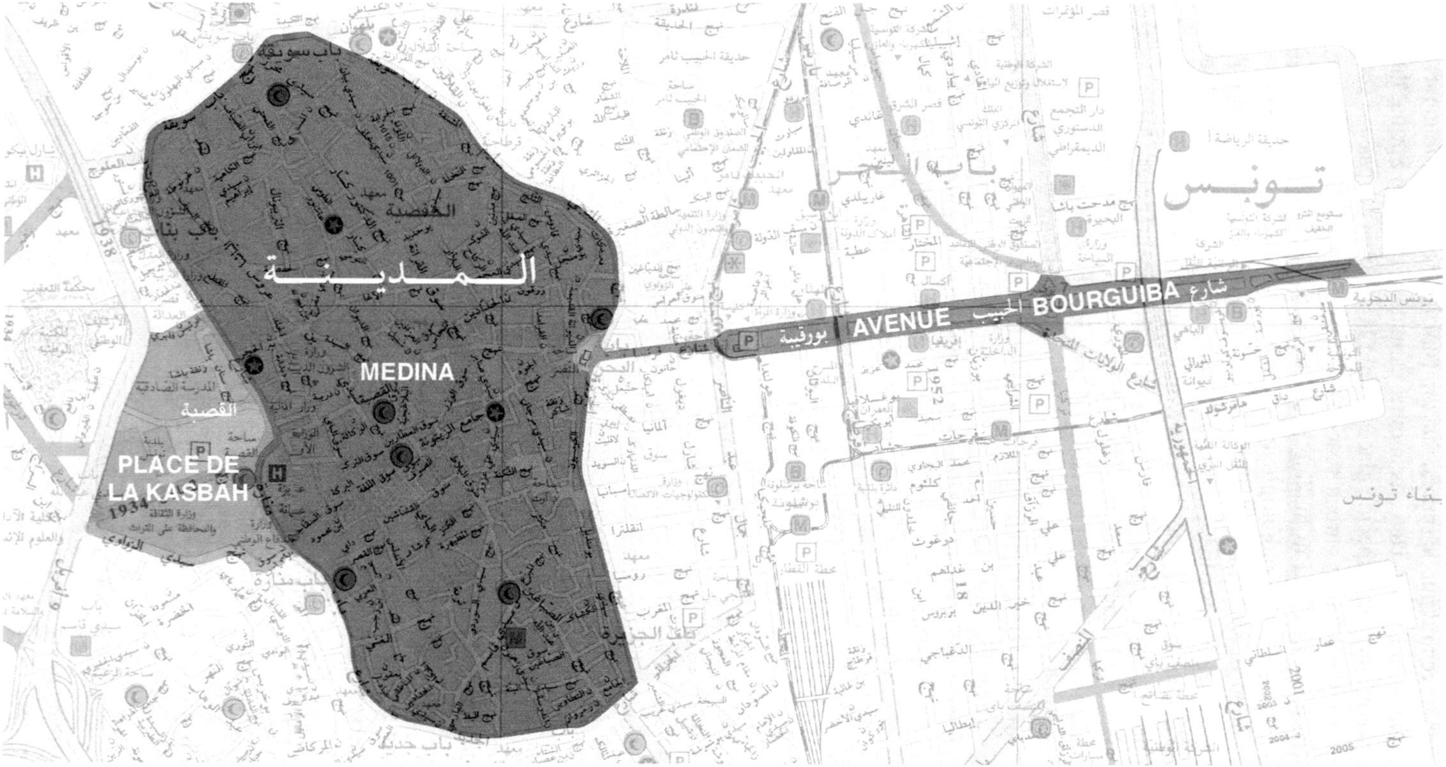

Figure 3.1 The three spaces of exploration within the Tunis's centre.

Source: Prospekt Photographers (www.prospektphoto.net), 2015.

the country and in the suburbs of the capital, it found here, in this wide boulevard, its final epicentre, its amplifier, and its spatial epiphany. Echoing Saskia Sassen (2011, p. 574), during the Revolution, Avenue Bourguiba acted as a global street, that is 'a space where new forms of the social and the political can be made, rather than a space for enacting ritualized routines'.

During the crucial day of 14 January 2011, it was here that the final face-off took place between the regime, symbolically besieged within the palace of the Ministry of Internal Affairs, and the counter-power represented by the thousands of protesters that asked for Ben Ali's resignation (on the cry of *Dégage*!). 'I was really excited seeing that the number of people didn't cease to grow. There were resolute people of every kind, younger and older, of the most different social classes, everyone with the same aim: to reach the Ministry of Internal Affairs', says Larbi Chouika, a political scientist at the University of Manouba and a socialist activist, who recalls being still touched by this.

The Ministry of Internal Affairs and the functions that it used to perform at the times of the regime are an integral part of the Avenue Bourguiba and of the city. Because of its imposing materiality, the palace is an explicit symbol of political power and control, and it constantly reminds everyone that the power is always here; it is visible and recognizable, and it has a physical and corporeal presence. In the memories of Tunisians, the building represents, more than any other, the harshest forms of politics under Ben Ali's regime: control, censorship, repression, violence, and so on. The power of the regime was embodied in this stark and tangible palace, but at the same time it pervaded, repressive and stifling, the streets of the city and people's lives.

In fact, the regime's power expressed itself not only in the interdiction of any form of public manifestation and contestation but also in the censorship of thoughts and discourses. Sport was actually the only topic one could freely speak about and upon which one could 'freely' share opinions with others. 'Once you could only say I'm *clubiste* or *esperantiste*, the two football teams of the capital. Most of all, we were used to chat about football, Milan or Barcelona [. . .], but we never debated politics, because you really had to pay attention', an artisan in the souk of the *chechia* remembers.

In antithesis to these recent memories, the Ministry stands nowadays as one of the symbols of the liberation of the country, as it bestows the sense of unity all Tunisians feel about the making of the Revolution. In this regard, Emna, a young graduate in architecture that we met in a café located in front of the palace, tells us: 'The mere sight of the Ministry is frightening because the most evident architectonical element of the palace is the fear it instils. We did not even dare to look at it [. . .]. Today we have changed the way in which we take possession of public space [. . .]. Now I feel that all this is ours. This is our space, our place, our street, our country'.

The austere and impersonal facade of the palace of the Ministry exemplifies its extraneousness with respect to the everyday practices that take place in the Avenue Bourguiba; the barbed wire that encircles it, with the aim of protecting the building, prevents anyone from approaching it, conferring a sense of self-isolation upon it. Two years after the Revolution, the barbed wire is a recurring element of the urban geography of Tunis, a true micro-infrastructure of the post-revolutionary city (Amin, 2014). Apart from enclosing the Ministry and the *trottoir* at the middle of the Avenue Bourguiba, it reappears at the entrance of the street in Place de l'Independence and in front of the French Embassy, all places that are permanently guarded by military deployments too.

The barbed wire is an urban element that is only apparently temporary and removable. In reality, it has been a constant presence since the times of the Revolution and seems to have acquired permanent or somehow lasting features. It is certainly an element for division and separation, deeply affecting the space, creating suspended and inaccessible places that have been stolen from the everyday practices and uses involving the remaining part of the city. Blocking some sections of the central promenade of the Avenue Bourguiba, it interrupts the rhythm of the movement of people and forces them to bypass it, dodge it, and pass it by without touching it, and to occupy the roadway to step round it anyway. From a more symbolic point of view, the presence of the barbed wire expresses the persistence of a form of control over public space that is intended to separate social and political bodies, emphasizing the reappearance of social tensions and violence, as well as the diffused sense of fragmentation within the country and the many difficulties of the democratic path.

The political meaning of the Ministry of Internal Affairs is in strict relation to another highly symbolic building on the Avenue Bourguiba, the Municipal Theatre. Positioned at the very centre of the street, the Theatre is one of the most evident legacies of the French colonial presence in Tunis, together with the close-by Saint Vincent Cathedral, the French Embassy, and the other liberty-style buildings along the boulevard and its cross streets.

Since the start of the Revolution, the Theatre has represented an elective place of social assemblage and aggregation, a physical and metaphorical heart of political identification and symbolic appropriation. Even more so now, the Theatre represents a political space, a natural stage, the occupation of which is claimed and contended by forces, movements, and practices that are radically antithetical.

However, the Municipal Theatre is also one of the most practiced and lived in public spaces within the centre of Tunis; it is the official epicentre of the artistic and cultural life of the capital, and it is the place where a dense knot of spatial practices finds its stage. The small square in front of the Theatre and the staircase at the entrance are the places people use to pause and transit, take a break, bump into each other, or arrange a meeting. At night, when the Avenue becomes less welcoming and more uncomfortable, the space around the Theatre remains a safe and lived space, where people stay and meet together and where groups of youngsters improvise rap performances or breakdance. Seated at the small tables of the contiguous Café de Paris, one of the most popular of the city, it is impossible not to see people at all times who are eat, taking tea, talking, or simply lingering.

In the geography of Tunis, the café seems strongly characterized by the class, age, provenance, gender, and the political belonging of its clientele. Now, after the Revolution, when the fear of the regime's censorship and the practice of informing has definitely passed, the café seems to have acquired that function of public space *par excellence*, as indicated by Habermas (1962). So far, cafés cannot be interpreted as spaces of integration because in people's attendance rigid forms of segregation (for example, between ages and genders) still stand. Anyway, the current practices within the cafés of the centre of Tunis can be reconnected to the Arab-Turkish tradition that sees them as a place of political discussion and confrontation (Chaouachi, 2002).

Just as during the Revolution, the Avenue Bourguiba still operates as a space of contestation, where people converge to make their social and political role visible and recognizable. People pour into the Avenue to demonstrate and to claim their rights and present their causes or to contest the political power. In this regard, Selma Jobbes, the owner of the Al Kitab bookshop at the very centre of the boulevard, states: 'Every time there's a demonstration, it has to be here, even if not every shopkeeper or police officer agrees that. All the same, in my opinion this is inevitable because the Avenue is now a historical place endowed with a very peculiar emotionality. Personally, I do not have any difficulties with these demonstrations. On the contrary, more are needed! They are the signal of the establishing of democracy'.

Along the Avenue, small or large, spontaneous or organized demonstrations, protests, and pickets happen almost every day. The women demonstrate to claim their rights and laicism; the oppositions shout out their anger after the assassination of Chokri Belaid (the leader of the left-secular Democratic Patriots' Movement, assassinated on 6 February 2013); at the beginning this was spontaneous, later on it was through publically coordinated demonstrations repeated every Wednesday, at the same time, in front of the Municipal Theatre. On 9 April, the

Figure 3.2 A showcase of the everyday life (Avenue Bourguiba).

Source: Samuele Pellecchia, Prospekt Photographers, 2013.

Avenue hosts a great manifestation in memory of the Martyrs for the Independence of 1938 (when the first uprisings against the French colonial power started the liberation of the country). Almost every political entity takes part in the event, facing each other and competing for the spaces in the Boulevard.

Hence, the Avenue Bourguiba is a showcase of social and political behaviours that, after the downfall of the regime, have acquire renovated and unedited meanings (Figure 3.2). Right inside the Avenue, newly politicized practices and claims mix with the ordinariness of the everyday social interactions. Here people meet; date; watch each other; stroll; window shop; and attend public concerts, events, and performances.

The square that changed Tunisia

As well as Avenue Bourguiba, the Place de la Kasbah power has also acquired specific spatial forms and meanings. However, the modes into which power is *spatialized* are quite different in the two cases. Within the Avenue Bourguiba, power spreads itself into the everyday life of the city, almost hidden within the frenetic rhythm of people, encounters, and trades. Opposite to this, the square of the Kasbah is the spatial manifestation of the command centre, where power safeguards itself through its separation from the rest of the city. Although Place de la Kasbah is actually a part of the centre of Tunis, at the same time it is placed outside it. Being somehow detached from the city, the Kasbah controls it and guards it from on high. After all, as the well-known Tunisian urban planner Jellah Abdelkafi says, 'the Kasbah is a fortress', and not just metaphorically: 'It's the place of the authority, all the political and the administrative power was there, aloft'.

In Arabic, Kasbah is the generic name indicating the citadel above the urban agglomeration, and it is commonly used, extensively, to include the entire historical urban pattern previous to the occupation. Originally, the term designated a military fortress, surrounded by defensive walls, which was separated from – or adjacent to – the rest of the city and commonly placed in a strategic and elevated location. In addition to its innate and intrinsic defensive function, with the passage of time, it acquired a function of control and domination over the city. The Kasbah still remains strictly connected to political power, to its *mise en scene*, to its legitimation and, as well, to its defeat.

While Avenue Bourguiba is a practiced and lived space, Place de la Kasbah is a performing stage on which the State's power and monumentality is celebrated, and that strongly hinders and inhibits the development of everyday practices.

In and around the square there are: the Palace of the Government; the headquarters of the Socialist Desturian Party founded by Habib Bourguiba (and today the headquarters of the Municipality of Tunis, where moreover nobody seems to go in or out); the palaces of the ministries of Defence, of Culture and Communications, of Religious Affairs, of Finance, of Justice, of Education, and that of Social Affairs; the palace of the Tekia, headquarters of the directorate of the Democratic Constitutionally Rally, the hegemonic political party founded by Ben Ali in 1988 and declared illegal in March 2011; the headquarters of the United Nations; the courthouse; the Aziza Othmana Hospital (once a military barrack); the Mosque of the Kasbah (which is currently closed); the Sadiki college, which is the first high school of 'modern' conception in Tunisia, where the most part of the capital's intelligentsia studied during and after the French protectorate; and, in the middle of the square, the National Monument to Independence that commemorates the Tunisian martyrs of 1938.

Even if the defeat of Ben Ali's regime was performed in the Avenue Bourguiba on 14 January 2011, it is thanks to the occupation of the Kasbah during the following weeks and months that the Revolution carried on, becoming a long-lasting event. After all, Avenue Bourguiba is the space of the rupture, which was violent and unexpected, while the Place the la Kasbah is the space of the duration and persistence. 'This square has changed Tunisia, that's for sure! This square and the people that were here, occupying it', recalls Yassine Sami Turki, urban planner at the *École Nationale d'Ingénieurs* of Tunis. 'I will never forget an episode I attended one late night during the occupation. There was a group of young people engaged in cleaning the space around and two girls were working harder than the others were. One was wearing the headscarf, the other was not [. . .], but they were working together, side by side. I remember I thought: This is Tunisia, it begins to take shape'.

Even now, this flat and blank square, designed to host large gatherings and celebrations, is determining the future of the country. Echoes and proofs of the occupation have faded away. The countless graffiti that had been pictured on the walls and on the pavement, photographed and shared worldwide over the Internet as evidence of the recovered freedom of speech and of artistic expression after the regime's censorship, have recently been cancelled and covered by the white paint of the post- or neo-revolutionary temporary order.

Nevertheless, the square is perceived to be one of the most symbolic spaces of the Revolution, and people go on demonstrating with the aim of finding legitimization, visibility, and recognition of their claims and rights. Within the space of Place de la Kasbah the most diverse representations, typologies, and practices of political power meet and clash together.

From the disputes between the rival political coalitions competing for supremacy over the government to the several interested micro-groups claiming a right to a small piece of street, from the control exercised by the police over the square to the various everyday uses of it, the space of Place de la Kasbah is innocent only in appearance and is actually undermining the sense of sacredness and inviolability of the Kasbah and introducing elements of ordinariness and normality.

Almost every day, the square hosts small or large demonstrations: an assembly of civil lawyers asks for a new statute; a group of nurses protest against the closure of a hospital; an assemblage of Islamists search for visibility; and so on. All of these practices, together with the incessant coming and going of taxis, the flow of tourists coming from the cruise ships, the boys who play football day and night, contribute to the *normalization* of the Kasbah as an open, public, and collective space. In their making, all of these spatial practices are revolutionary if compared

Figure 3.3 Place de la Kasbah as a suspended space.

Source: Samuele Pellecchia, Prospekt Photographers, 2013.

to the inflexibility and formality of the prohibitions that the regime guaranteed (and that guaranteed the regime itself) until 2011.

Traversed by this amalgam of meanings and practices, the Kasbah appears nowadays as a suspended space, waiting for something o a place that does not belongs to anyone, and thus belongs to everyone (Figure 3.3).

The memory of the city

The fluid and shimmering combination of practices taking place along the Avenue Bourguiba and the different roles that space plays within the city are impossible to understand if one does not walk through the Medina. Almost naturally, the wide line of the Avenue is dialectically connected to the historical and original space of the Medina. The functionaries of the French Protectorate used to describe this space as *irrationnel* and *déraisonnable* (Berque, 1958), as a physical impediment to modern and rational urbanization, and, even more symbolically, to the progress itself that should have been introduced by the paternalistic colonizer (Troin, 1997). Many descriptions and readings of the southern Mediterranean city start from this same assumption. On the one side, the Arabic/Muslim/Islamic/eastern city with its irregular spatiality (Serjeant, 1980; Boudhiba and Chevallier, 1982; De Epalza, 1991); on the other side, the colonial city and the rationality of its Cartesian/orthogonal city plan, designed in a way that evokes the Hippodamian

Figure 3.4 The threshold of Bab al-Bahr – Porte de France.

Source: Samuele Pellecchia, Prospekt Photographers, 2013.

grid system, which should operate as a regulator of democratic and egalitarian practices (Rossi, 1987).

The simultaneous separation and connection between the Medina and Avenue Bourguiba, between the historical and the colonial city, occur in Bab-al-Bahr (Porte de France), one of the many gateways of the Medina of Tunis. Bab al-Bahr not only divides and connects the Medina to the colonial city, the space of *tradition* and (the myth of) *modernity* (Figure 3.4).

It also stands between the spaces of the Revolution and the most historical, identitarian, and symbolic places of the city. The borderline between the two is both physical and emotional. The Medina is the heart and the memory of Tunis, and Avenue Bourguiba is its modern counterbalance, both from the morphological and ideological point of view. The practices and the symbolic values that fill these two spaces are blended together in a very complex relationship that cannot be confined to the reductive distinction between antique and modern, tradition and novelty.

During its long history, the Medina of Tunis has come under many revolts (Abdelkafi, 1989). Nevertheless, it did not play a central role during the Revolution of 2011. Despite its inner position between the Avenue Bourguiba and the Kasbah and its geographical centrality, the Medina appears to be a residual space, which has been preserved and left apart from the competition for taking possession (in real or symbolic terms) of the other public spaces of Tunis. Thus, the Medina is a defended space, as is the historical heritage it contains. Somehow,

Figure 3.5 Memory and identity: the Medina as space apart.

Source: Samuele Pellecchia, Prospekt Photographers, 2013.

it remains distant from the other spaces of the city where political movements, activists, militants (either liberals, conservative, or radicals), workers of every kind (lawyers, teachers, policemen, or artists), former political prisoners, religious and non-religious groups, youngsters, and women and children have claimed their right to existence and recognizance.

At this time of rediscovered freedom, the Medina stands apart (Figure 3.5).

Its identity and diversity, its memories and traditions find protection and shelter in the daily practices that take place in between. 'Every day I open my shop and it's always the same. One day it works, the other does not. Life is like this', Amor affirms. Since 1970, he has run a carpet store in the core of the Medina. Here, even after a Revolution, things do not change: 'It's normal. Everything changes, there's freedom, but the souk always remains the souk'.

During the daytime, the Medina takes its life from the chaos of the market and the many artisanal activities, in the sounds of the crowd that stop and go by, slipping through its narrow streets, in the movement of people buying or swapping or who sit in a café or enter a mosque. As night falls, the souk and the café rapidly empty and the gateways encircling the central square around the Jamaa Ez-Zitouna Mosquee are fastened. Silence spreads into the whole Medina, which becomes a dense space, hidden and apparently empty, guarded and defended by men armed with wooden beams, who are custodians not only of the space but also of a treasure, of a memory, of the whole city, of its history, of its persistent order, of its power, of its future.

The revolutionary practice of rewriting Tunis's public spaces

During the Arab Spring, claims, protests, and demonstrations have reaffirmed the role of urban public spaces as centres of conflict and of politics, spaces of political expression and reworking, and of emancipation and solidarity. The public spaces in the centre of Tunis are not simple containers of revolutionary movements and moments, but they have played at the same time the role of *enjeu spatial*, places of social and political practices, physical and symbolic epicentres, and infrastructures of and for change.

Two years after the Revolution, the legacy of the multiple roles played by space persists. The practices of life in the central places of the city have settled into lasting forms of reappropriation of public spaces as spaces of expression (political and otherwise), of claims, debate, discussion, and action. The plural potentials of urban practices as generative of political action keep over time 'the varied role that urban processes sometimes play in generating, recognizing, mediating, circulating and addressing political claims' (Barnett, 2014, p. 1639). Subjects who were first politically and socially excluded, then marginalized and made invisible by censorship and repression imposed by the regime, regained the city, defining and affirming its role as a *space of appearance*. Claims are now visible and recognizable, as expressed in the political and spatial change. According to Swyngedouw (2001, p. 3), '[S]uch political events are interventions that transgress the symbolic order and mark a shift to a new situation that can no longer be thought in terms of the old symbolic framing. [. . .] Such new symbolizations are where a possible repoliticization of public civic space resides. It emerges where those who are un-counted and unnamed, whose fantasies are only registered as noise, produce their metaphorical and material space'.

In Tunis, two years after the Revolution, fantasies which are no more mere noise, but which express and produce their space, are those of the kids playing soccer in the Place de la Kasbah, of the people who walk freely and talk of politics in café, of the merchant of the souk of *chechie* who, looking into the camera, says, 'Vive la Révolution!'

'On ne reviennent pas en arrière'. We have heard these words repeatedly: fundamental freedoms, denied during the regime (freedom of speech and expression), and now achieved, may no longer be questioned. Even the legislative and presidential elections of the autumn and winter of 2014, which took place in an atmosphere of relative calm and of political and social accountability, seem to confirm this confidence in the future.[3] Tunisia appears to be an almost unique case among the countries affected by the Arab Spring, which are now passing through a dramatic and labourious process of transition, if not of restoration of new forms of regime. Of course, the date of 18 March 2015 casts a shadow over the future of Tunisian democracy, but not over the spatial and social changes.

It is a change that is expressed in the public space that appears as though it is an open social laboratory, where different spatial practices meet (and sometimes clash) in a process of constant redefinition, negotiation, and affirmation of values, rights, belonging, and uses. Demonstrations, pickets, presidia, flash-mobs, occupations, and debates, but also simply the talk in the cafes, the ways of appearing in public and relating to each other all become micro-revolutionary practices, constantly renewing their timing and modes of attendance, constantly changing form and the social, political, and cultural meanings of the urbanity and of *being in public*.[4] These practices and awarenesses assume a revolutionary dimension because they give rise to a reconfiguration of political and civil being, of the production of discourse, and of practices of mobilization that had been impossible in Tunis for many decades.

The occupation of urban space through demonstrations and protests, as well as the sedimentation of change expressed through everyday acts and practices, play a major role in political action. According to Davidson and Iveson (2014, p. 142) and echoing Jacques Rancière, occupation of urban space is 'neither political because of its location in a pre-designed political or public space, or because it involves the gathering of demand-making people in a given space. Rather, the politics of an occupation is a question of subjectification and reinscription – that is, occupations become political to the extent that they involve the subjectification of a "part of those with no part", a group that challenges the part supposed to be played by the bodies in that particular place in the process of declaring their equality'.

Within this constant *becoming* and *overlapping* of practices and meanings, to find fixed points is difficult and perhaps not even necessary. It is impossible to wonder if Tunisia is now more democratic, more just, more equal (or safer, richer, more secular, etc.) than prior to the Revolution; or whether the Revolution has brought more rights and opportunities to the most vulnerable or marginalized social groups. Nor, again, is it possible to assess whether these values and parameters are correct for judging the Revolution. Tunisia still is a country crossed by deep social and economic inequalities, and, already in the spring of 2013, we could feel some disillusion about the concrete changes that the Revolution could have produced.

However, echoing the notion of the *virtual object* of Henri Lefebvre, Purcell (2013, p. 23) sees democracy as 'a horizon toward which we can move'. As with any horizon, even the horizon of democracy is moving and we cannot reach it. Therefore, *doing democracy* is not to construct a formalized political order but an active and perpetual engagement 'to move politically from oligarchy to democracy, from passivity to activity, from heteronomy to autonomy' (*ibidem*, p. 8), from injustice to justice, and from marginality and exclusion to participation and inclusion. In public spaces in the centre of Tunis, the movement towards the virtual object of democracy and towards the horizon of justice is a process of constant redefinition, negotiation, and affirmation, and it is impossible to fix and represent it once and for all. As any spatial phenomenon and process (Thrift, 2008), post-revolutionary Tunis cannot be fully described or represented. Although 'the world is so textured as to exceed our capacity to understand it' (Davies and Dwyer, 2007, p. 258), showing the spatial practices gives political and poetic value to the changing multiplicity of the *geography of what happens*. This value expresses itself through the continuous redefinition of marginality and privilege, of claim and conflict, of the visibility and invisibility of subjects and spaces.

Notes

1 On 18 March 2015, while we were writing this chapter, a terrorist attack with many victims bloodied Tunis, challenging the difficult and uncertain path towards democracy. Obviously, we do not know what will happen, as it would not make sense to put forward hypothesis or interpretations. This text is a small tribute to the Tunisian Revolution, to the people we met in Tunis, to their strength, hopes, and dreams.

2 The postdisciplinary research group involved in the field study included eight people: four geographers – Raffaele Cattedra, Maurizio Memoli and Matteo Puttilli (*Università di Cagliari*), and Francesca Governa (*Politecnico di Torino*); three photographers and filmmakers – Bruno Chiaravallotti, Claudio Jampaglia, and Samuele Pellecchia (*Prospekt Photographers, Milano*); and, the freelance photographer Rosi Giua. For a more detailed presentation of the results of the research, as well as of the explorative methodology adopted, see the website (webdoc.unica.it).

3 The legislative elections were held on 26 October 2014, and the presidential election was on 21 December 2014. Beji Caid Essebsi, founder and leader of the majority party of the secular matrix Nidaa Tounes, won the ballot.

4 According to some women interviewed during the fieldwork, the Revolution has defeated the prohibition to wear the headscarf in public places imposed by Ben Ali's secularist-biased regime. For many women, wearing the headscarf therefore acquires a new meaning: It is certainly worn as a religious symbol of identity and membership but also as an act of asserting one's own freedom of expression and religion.

References

Abdelkafi, J., 1989. *La Medina de Tunis. Espace historique*. Tunis: Alif.

Amin, A., 2014. Lively infrastructure. *Theory Culture & Society*, 31, 7-8, pp. 137–161.

Barnett, C., 2014. What do cities have to do with democracy. *International Journal of Urban and Regional Research*, 38, 5, pp. 1625–1643.

Berque, J., 1958. Médinas, villeneuves et bidonville. *Les Cahier de Tunisie*, 6, pp. 5–42.

Bettaïeb, V., 2011. *Dégage. La révolution tunisienne, 17 décembre 2010–14 janvier 2011*. Paris: Editions du Layeur.

Boudhiba, A. and Chevallier, D., 1982. *La ville arabe dans l'Islam. Histoire et mutations*. Paris and Tunis: Ceres-Cnrs.

Brenner, N., Marcuse, P. and Mayer, M., 2012. *Cities for people not for profit*. London and New York: Routledge.

Butler, J., 2011. Bodies in alliance and the politics of the street (online). Available at: http://www.eipcp.net/transversal/1011/butler/en [accessed 26 Apr. 2014].

Chaouachi, K., 2002. *Le monde du narguilé*. Paris: Maisonneuve.

Davidson, M. and Iveson, K., 2014. Occupations, mediations, subjectifications: fabricating politics. *Space and Polity*, 18(2), pp. 137–152.

Davies, G. and Dwyer, C., 2007. Qualitative methods: are you enchanted or are you alienated? *Progress in Human Geography*, 31, 2, pp. 257–266.

De Epalza, M., 1991, *La ciudad islámica*. Saragozza: Institución Fernando el Católico.

Deleuze, G. and Guattari, F., 1972, *L'anti-Oedipe. Capitalisme et schizophrénie*. Paris: Minuit.

Filiu, J.P., 2011. *The Arab revolution: Ten lessons from the democratic upraising*. London: Hurst.

Gobe, E., 2012. Dossier: Un printemps arabe?, *L'Année du Maghreb*, VIII.

Gregory, D., 2013. Tahrir: politics, publics and performances of space. *Middle East Critique*, 22(3), pp. 235–246.

Habermas, J., 1962. *The Structural Transformation of the Public Sphere: An Inquiry into a category of Bourgeois Society*. Cambridge: Polity.

Harvey, D., 2013. *Rebel cities: from the right to the city to the urban revolution*. London: Verso.

Kanna, A., 2012. Urban praxis and the Arab Spring. *City*, 16(3), pp. 360–368.

Lefebvre, H., 1970. *La révolution urbaine*. Paris: Gallimard.

Lefebvre, H., 1974. *La production de l'espace*. Paris: Anthropos.

Lipietz, B. and Lopes de Souza, M., 2012. Introduction: where do we stand? New hopes, frustration and open wounds in Arab cities. *City*, 16(3), pp. 355–359.

Lopes de Souza, M. and Lipietz, B., 2011. The Arab Spring and the city. Hopes, contradictions and spatiality. *City*, 15(6), pp. 618–624.

Mabrouk, M., 2011. A revolution for dignity and freedom: preliminary observations on the social and cultural background to the Tunisian revolution", *The Journal of North Africa Studies*, 16, 4, pp. 625–635.

Magnusson, W., 2014. The symbiosis of the urban and the political. *International Journal of Urban and Regional Research*, 38, 5, pp. 1561–1575.

Purcell, M., 2013. *The down-deep delight of democracy*. Oxford: Wiley-Blackwell.

Rossi, P., 1987. *Modelli di città. Strutture e funzioni politiche*. Torino: Einaudi.

Sassen, S., 2011. The global street: making the political. *Globalizations*, 8(5), pp. 573–579.

Serjeant, R.B., 1980. *The Islamic city*. Paris: Unesco.

Swyngedouw, E., 2001. Every revolution has its square. *cities@manchester*, (blog) 17 Jun. Available at: http://citiesmcr.wordpress.com/2011/03/18/every-revolution-has-its-square/ [accessed 18 Jun. 2015].

Thrift, N., 2008. *Non-representational theory. Space, politics, affects.* London: Routledge.
Troin, J-F., 1997. Il suk e il bazar: quale avvenire per la città della storia? In: AA.VV., *Città e società nel mondo arabo contemporaneo. Dinamiche urbane e cambiamento sociale.* Torino: Fondazione Agnelli, pp. 145–162.
Vasudevan, A., 2014. The autonomous city: Towards a critical geography of occupation. *Progress in Human Geography*, 39, pp. 1–22.
Weslati, S., 2011. *Démocratie ou guerre civile: Chronologie de la révolution tunisienne.* Tunis: Nirvana.

4 Tasty vehicles

Gourmet taco trucks, 'trap' parks, and other planning for *zombis fresas* (wealthy zombies) in San Antonio, Texas

Mark Tirpak

Introduction

My aim with this chapter is to help bring to light new formal, or government and select private partner, place-(re)making efforts in downtown San Antonio, Texas, that appear to burden many residents and visitors in this major but often overlooked US city, including by challenging some established local norms and material practices related to mobile food vending and the use of municipal parks.

A longstanding majority '*Latin@*' (Demby, 2013) and predominantly *Mexican@* or *Tejan@* city growing by birth as well as diverse migration from Mexico and elsewhere, San Antonio is an understood demographic bellwether for the US (Dotan, 2012; Cave, 2014); San Antonio has also been identified as one of the nation's most economically segregated major urban areas, with persistent and in some cases worsening poverty in neighbourhoods surrounding the city's intertwined downtown business, tourism, and 'public improvement' districts (Olivo, 2013; Sanders, 2014; Casura, 2015). Recent formal place-making efforts related to an ongoing 'Decade of Downtown' in San Antonio (Rivard, 2014) appear to be shaped by historical and current local, national, and international patterns of urban economic segregation, including increasing 'skyboxification' or economic stratification and parallelism in the US (Morello and Melinik, 2013) and class divisions in contemporary Mexico (Johnson, 2013; Rueda, 2014; Miller, 2015) and elsewhere. Following national and global trends, recent formal place making in downtown San Antonio has included new 'pop-up' style events and offerings incorporating mobile food truck and trailer vending, *Latin@* inspired and led and otherwise. These practices and services, while borrowing from more informal and improvised urban infrastructures prevalent if not originating in the city's working poor and predominantly Mexican@ or Tejan@ neighbourhoods, appear in San Antonio to be part of new formal efforts to deter various initiatives by and opportunities for the city's poorest or most frugal residents and visitors.

To add to recent literature exploring widening socio-economic inequality, transforming 'poverty politics' (Elwood, Lawson, and Nowak, 2015, p. 124) and related emerging 'defensive' or exclusionary place-making structures and practices in the US (Andreou, 2015) with a timely, site specific, and more critical and 'auto' ethnographic accounting, I focus with this chapter towards formal place-making *movidas*, or structured scenes and individual maneuvers (Spener, 2010), at and near the recently 'revitalized' historic Travis Park in downtown San Antonio (Parker, 2014). Coopting the imaginary of a 'City on the Rise' (Sanders, 2014) that has been championed by at least some downtown San Antonio boosters in recent months, I contend that some investments and practices at Travis Park, including new niche events, outdoor furniture installations, and use of the historic park as a site for the City's experimental Downtown Food Truck Program (City of San Antonio, 2014), might best be perceived as being directed towards enticing, satiating, and farming imaginary *zombis fresas* – wealthy, mindless, sequestration-(and somehow not toilet-)requiring 'consuming-machine' visitants (Lauro and Embry, 2008, p. 99) – at the expense of many actual existing San Antonio residents and visitors. I do so with hope that this more fanciful reframing of some formal place-making efforts in downtown San Antonio might help to broaden current local discussions about 'gentrification' and public improvement (Davila, 2015), as well as encourage new local responses to planned or publicly supported economic segregation or other vulnerabilities for a large and diverse population of the city.

I begin this chapter by recounting a recent and relatively expensive 'street food' meal of tacos and a soda served by a City-selected and closely monitored Downtown Food Truck Program vendor at Travis Park.

No glass 'mexi' bottles for ciudand@s (the people of the city)

The food truck vendor appeared embarrassed and somewhat nervous as he remembered the rules of vending at Travis Park, one of the oldest municipal parks in the US (Parker, 2014) and located just a few blocks from the famed Alamo as well as an international bus station where a coach from Mexico had just arrived. 'I can't let you take the soda bottle or the police will fine me 500 dollars', he said,[1] nodding towards a uniformed officer in the park behind me. As I glanced at the officer and noticed a police SUV parked a few spots up the street from the lone food truck, the vendor poured most, but not all, of the purchased and opened glass bottle of a citrus-flavoured soda imported from Mexico into a translucent plastic cup to complete my order.

I balanced on top of the flimsy cup a paper plate containing five *tacos al pastor*, spit-roasted pork 'mini' or 'street' tacos. *Tacos al pastor*, a historical hybrid of Lebanese and Mexican cuisines (Carman, 2009), is currently a common dish in San Antonio – so much so that it has served as a judging item for new corporate-sponsored taco truck 'competitions' to have emerged in the city in recent years (McInnis, 2014), as part of the understood *barrio*, or working poor Mexican-American neighbourhood, inspired national and now global food truck movement and phenomenon (Gold, 2012). Carefully, I made my way towards one of a number of brightly hued café tables placed near the centre of Travis Park and overlooked by a portable mini-library/games kiosk; these items, along with a giant chess set, form some of the new moveable and securable furniture comprising part of the recent public-private $500,000USD 'makeover' and 'rebranding' of the park (Bailey, 2014). *En route* to an empty table, I passed in Travis Park the pairing

Figure 4.1 New recycling bin (suggesting that glass bottles are permitted) at Travis Park.
Source: Mark Tirpak, 2014.

of a new permanent City trash can and separate recycling bin designed and labeled for, amongst other items, 'glass bottles' (see Figure 4.1).

It was just after 1:30p.m. on a weekday in early summer, and I was alone in Travis Park except for the police officer, a few City Parks workers tending to new decorative landscaping, and the food truck parked kerbside, operating as a private business 'participant' in the City's recently established Downtown Food Truck Program. As described in published rules and regulations, the City endeavours with the new programme to 'monitor and control the quantity and quality of vendors who desire to operate a mobile food truck vending operation' in downtown San Antonio, in order to 'create a sense of place' and 'provide a vibrant culinary experience at street level for downtown'. These objectives allude to the popularity and near-ubiquity of mobile food vending, including truck- and trailer-based vendors, in at least sections of San Antonio currently and historically – particularly, local approaches to selling 'Mexican' street foods downtown and in surrounding neighbourhoods, as shaped by local and migrating tastes, existing urban infrastructures, and other factors (Arellano, 2012; Pilcher, 2012). More critically, the Downtown Food Truck Program places City staff members in the unusual role of selecting and monitoring, if not curating and managing, more exclusive or intentionally rarified mobile street food dining 'experiences' for downtown customers in select parks and plazas, beyond simply regulating independent food truck and trailer operation based on established or new public health, safety, and business rules and licensing.

A key aspect of San Antonio's new Downtown Food Truck Program appears to involve City staff members reviewing proposed menus and selecting vendors depending on 'the truck's appearance, food quality and variety (avoidance of food redundancy)' with no stated guidelines as to how these criteria might be interpreted or judged by the City. Also absent from the published rules is any indication that the City of San Antonio aims to improve access to affordable and nutritious meals via this food-related downtown programme. It is a surprising oversight considering that many Downtown Food Truck Program sites also serve as stops and transfer points for public buses serving neighbourhoods throughout San Antonio, including areas of persistent and growing poverty and understood food insecurity near the downtown area (Everett, 2012; Sanders, 2014).

Related, there is no indication that the menus of the City's downtown food trucks and trailers are evaluated or vendors chosen based on their provision of items meeting *¡Por Vida!* nutritional standards, a City-County public health project to promote healthy menu options that currently targets ' "mom and pop" Mexican restaurants' (Robert Woods Johnson Foundation, 2012) and other prepared food vending practices in neighbourhoods within a few miles of Travis Park and other Downtown Food Truck Program sites. Through the Downtown Food Truck Program, the City also does not appear to help address the disproportionate share of poverty wage employment to be found in downtown San Antonio currently, including some hotel work (Sanders, 2015) and other public-facing and tourism-related positions such as the uniformed downtown '*Amigos*' visitor ambassadors and street cleaners (Olivo, 2013), by requiring that participating food truck vendors and their employees be paid at least the City's and the County's established living wage of more than $11USD an hour (Gonzalez, 2014).

As I sat down with my taco plate and soda cup, a City Parks worker offered to open a table umbrella for me, and I asked about the absence of other park-goers this particular afternoon. He suggested the rainy weather but emphasized the pre-revitalization uses of the park as factors. 'It used to be bad here with drugs', he said. 'We are trying to give this park back to the people of the city'.

The people of the city, or *ciudadan@s* (Diaz and Torres, 2012, p. 22). I consumed my $12USD 'street' taco meal considering the apparent divisions between past and desired users of Travis Park, historically a productive space of orchard and other farming, but in the present defined primarily by City-deployed and -controlled portable place-making props and minimal actual public facilities; noticeable particularly to a diner, there are currently no permanent public toilets or even hand washing basins at Travis Park. As I sat and ate, I also thought about more generous and less expensive offerings of this fairly standard local combination of pork mini tacos and a Mexican soda, served from trucks and trailers as well as at various chain and 'mom and pop' eateries throughout San Antonio. Typically and noting San Antonio's history and current reality as a city with high levels of poverty-related food insecurity (Everett, 2012; Henderson, 2015), it is a meal served with the entire glass bottle of soda provided to the patron. This includes as it is an order more often than not 'stretched' amongst customers or shared with kids by passing the soda bottle and using the customary second corn tortilla of each double-wrapped mini taco to 'create' additional tacos.

Figure 4.2 Posted rules (which do not state that glass containers are prohibited) at Travis Park.
Source: Mark Tirpak, 2014.

In other parts of the US and abroad, Mexican sodas served in glass bottles have gained customers as part of the global 'gourmet' food truck movement and migration of some Mexican foodways; despite the commonplaceness of 'Mexi' sodas (Wong, 2013) in San Antonio as well as their association with novel and arguably more healthful consumption elsewhere, glass containers including soda bottles are currently forbidden by municipal ordinance at Travis Park, although various downtown parks and public spaces have been exempted from this ban (City of San Antonio, n.d.). While the glass container ban is not listed in new signage stating the 'rules' at Travis Park (see Figure 4.2) and appears to conflict with the City's recent recycling efforts at the park (see Figure 4.1), police enforcement of the ban has apparently disciplined at least some vendors as well as customers participating in the City's Downtown Food Truck Program at this site.

At least, the threat of enforcement this particular day left me dwelling on my purchased glass soda bottle and the remaining drink apparently trapped within the City's select taco truck by an odd expansion of 'quality of life' policing (Goldstein, 2014) – potentially, a new *botellas rotas*, or broken bottles, theory to justify continual police presence and interventions at Travis Park, but nonetheless marking a significant break from established local street food vending and dining norms, as well as challenging more frugal or imaginative consumption habits in San Antonio.

Displacing urban '*rasquachismo*'

In the predominantly working-poor and food-insecure neighbourhoods immediately surrounding downtown San Antonio and not far from Travis Park (Everett,

2012), an empty glass soda bottle likely has a different future or futures than being placed in a City recycling bin or put to nefarious uses, as likely envisioned with the City's glass container ban in select parks and public spaces. In these areas, it is easy to imagine or even spot glass soda bottles being 'upcycled' as part of any number of everyday creative or innovative efforts, from serving as the base for a new household item to functioning as an art or building supply. In the 'revitalized' but this day nearly empty Travis Park, such *rasquachismo* (Ybarra-Frausto, 1991) – or the local sensibility of making the most with the materials at hand, as informed by lean times and acquired aversions to waste – seemed out of the question if not 'illegal'. In contrast to a glass bottle, the pliable plastic cup, barely functioning as a drink container in its first use, felt destined for one of the park's new bins. At least, without the glass soda bottle to sip from or save and as the lone food truck closed operations for the day promptly at 2:00p.m., as mandated by City rules and monitored by the police officer standing by, every signal indicated it was time for me to move on until the next scheduled feeding or event.

As I exited Travis Park, I noticed close to the street an ominously gated new small confinement (see Figure 4.3). Identified as an 'off-leash' area for dogs, the permanent, single-entrance enclosure has fencing more than 5.5 feet (170 cm) high, or above and near the average height of women and men respectively in the US.

Confronted by this new cage-like public furniture and in growing need of a toilet, I left Travis Park pondering what kinds of *ciudand@s* or encounters were being planned for with this and other recent downtown 'improvements' and expenditures.

Figure 4.3 New excessively fenced permanent 'dog park' enclosure at Travis Park.

Source: Mark Tirpak, 2014.

Zombis fresas versus San Antonio's 'working poor' and 'threadbare'

While alluding to the reported influx of self-described *migrantes 'fresas'*, or ultra-wealthy 'strawberry' immigrants, from Mexico to San Antonio and elsewhere in the US in recent years through new economically defined visa pathways (Cave, 2014), I use the neologism *zombis fresas* to describe an imaginary body or swarm of wealthy, mindless, sequestration-requiring 'consuming-machine' visitants (Lauro and Embry, 2008, p. 99) of no particular background that appear to be planned for or summoned with some recent formal place-making efforts in downtown San Antonio. I use the term with respect for and a desire to expand San Antonio 'Spanglish' and multilingualism (Villa, 2012), as well as with awareness of how historic and evolving Mexican and Mexican-American class- and ancestry-based denigrations and sometimes self-identifications, from *'fresa'* to *'nac@'*, *'poch@'*, *'chunti'*, and *'junior'* (Arellano, 2005; Johnson, 2013), might complicate or challenge currently mainstream or more monolingual understandings of urban marginalization in the US. Particularly, as the US continues to grow as a Spanish-speaking nation, now second only to Mexico in number of Spanish-speaking residents (Burgen, 2015), and as extreme and worsening economic inequality and segregation in both Mexico and the US creates increasingly comparable and interwoven national experiences (El Daily Post, 2015).

Distinct from the decomposing and violently voracious 'cinematic' zombie (Lauro and Embry, 2008, p. 97) and equally removed from the smiling *Día de los Muertos*, or Day of the Dead, 'sugar skulls' and skeletons found in San Antonio's tourist markets year round (Blaze, 2014), the masses of oddly posed, minimally dimensioned, and otherwise uncanny humanoid figures milling about in various recent architectural renderings developed with new design software inform my San Antonio *zombi* analogy; this includes with awareness of the use of *zombi fresa*–populated computer-generated images to promote and justify various new projects and public investments in downtown San Antonio currently. Artificial images that are occasionally juxtaposed with photos of actual current users of targeted properties and areas, including Travis Park, engaged in seemingly desired public activities such as sitting and conversing but labeled as 'vagrants' or otherwise deviant (Rivard, 2013).

Such practices help to reveal what Ash Amin (2013) has described as the 'politics of severed society and space' (p. 438) as well as the 'cleansing class' (p. 489) sensibilities underpinning much current urban development and formal place-making work; they also help to map some of the 'brain' or dominant virtual as well as physical 'thinking' spaces and networks (Gandy, 2005, p. 32) of San Antonio at present. For some municipalities, building and managing new 'nodes of hyper-connectivity and premium service provision' within a 'wider landscape of neglect' has marked a transition to 'a new logic of differentiation and exclusion' (*ibidem*, p. 37). In contrast, recent formal place-making efforts in downtown San Antonio appear to draw from a deep well of local historical experiences with economic segregation as well as more contemporary 'differentiation' logics, including new quality-of-life policing approaches and more exclusionary mobile public dining practices that appear to have migrated to the city from elsewhere in the US, Mexico, and further afield.

One example is the recent arrival of the originally Paris-based *dîner en blanc* series of unexpected, if not secret and purposefully exclusive, meals at Travis Park, where an ostensibly 'free' public dining event was held to commemorate the first anniversary of the municipal park's revitalization and rebranding; attending guests, clad all in white and informed primarily by social media, were expected to supply their own meal or order a picnic basket in advance, in addition to reserving a park table with a minimum donation of $200USD (Greer, 2015). The intentionally minimally attended dinner, introduced to the gathered public as a new annual fundraising event for Travis Park, was supported by a handful of selected food truck vendors as well as a sizeable contingency of downtown district *Amigos* workers dressed in white *quayaberas*, uniformed police officers, and white-clad City staff members directing and serving guests. Although event organizers and staff members made efforts to invite individuals unaware of the dinner and passing through Travis Park to join the festivities on a 'first come, first served' basis following the seating of donors at tables, the invitations felt decidedly empty and insensitive not only to obvious potential barriers to participation but to the ongoing role of Travis Park in memorializing uncritically the US Confederacy; as it were, a purposefully obscure and confronting 'dinner in white' was held in the shadow of a statue at Travis Park commemorating the US Civil War's 'Confederate dead', with no additional signage to explain the long-controversial historical marker and despite recent major public investments in the park. Within weeks of this unusual public dining and fundraising event at Travis Park, police officers citing 'Downtown Mobile Food Truck' rules and regulations issued a $2,000USD fine against a licensed food handler and organizer of a less lavishly catered but longer-established mobile effort in San Antonio that provides weekly free 'restaurant-quality' and nutritious meals to anyone needing support, including homeless individuals, at a neighbouring downtown municipal park (Garcia, 2015). A park which, based on the understood 'success' of Travis Park revitalization in recent months and surrounding new private developments, now appears to be targeted for similar formal place-making efforts.

Echoing past progressive eras in San Antonio, 'hygenist discourses' (Gandy, 2005, p. 32) of health, vitality, and exposure have been employed to target some mobile public dining and feeding practices as well as to transform more improvised and independent street food vending efforts into exclusive and City-managed offerings for the benefit of imagined or desired patrons. With awareness of the long history of San Antonio's 'Chili Queen' vendors, traditionally an informal and open air vending practice engaging a diversity of residents and visitors in downtown streets and plazas but narrowed and ultimately ended by forced enclosure in City structures designed primarily for the perceived well-being of tourists (Pilcher, 2012, pp. 124–129), the likely results of the City of San Antonio's current downtown formal place-making *movida*s seemingly catering to *zombis fresas* come into view.

A municipal 'trap' park

Intimating that some recent efforts at Travis Park and surrounding downtown sites appear designed for non-human *zombis fresas* allows for an atypical critique of decidedly 'unpleasant' (Savičić and Savić, 2012) or hostile and marginalizing formal place-making structures and practices to have emerged in downtown San Antonio in recent months. This includes the 'dog park' cage and potentially entrapping glass recycling bins installed recently at Travis Park (see Figures 4.1 and 4.2), as well as the sudden removal of some public benches in the surrounding downtown area as a surprise policing tactic meant to discourage panhandling and loitering, as the need for food assistance and other support has apparently grown across the city (Garcia, 2015). As the number of homeless families in San Antonio increased a reported 19% and approximately 38% of requests for emergency food assistance went unmet locally in 2014, with a large percentage of the city's homeless and those requesting food aid currently employed (Henderson, 2015), the San Antonio Police Department in recent months petitioned the City to make the act of giving money, food, or other material aid in public to those asking for assistance a criminal offence (Garcia, 2015). Such policing efforts appear aimed at conditioning the public towards excluding or treating as suspect those exhibiting signs of poverty within the downtown area, as well as encouraging individuals of various means to limit the amount of time they spend and associations they make in publicly accessible spaces including parks. If a visibly poor, or simply dirty from work or travel, resident or visitor traversing Travis Park on the evening of the recent *diner en blanc* (Greer, 2015) had accepted the offer to join the festivities and grab an open seat at a donor table, depending on police enforcement but based on current trends, that individual likely could have been booked on any number of charges related to curbing 'aggressive' panhandling in the downtown area (Garcia, 2015).

In addition to challenging urban ethnographic research, where a casual conversation with a stranger in a park or on the street can prompt a police intervention related to new anti-panhandling enforcement, increasingly invasive quality-of-life policing efforts in downtown San Antonio including the removal of some public benches have steered some physically encumbered residents and visitors towards interfering with pedestrian traffic, another finable or arrestable offence depending on the perpetrator (Tedesco, 2014). Such inhumane ploys bring to mind the growing popular use of the word 'trap', as originated in the US South and spreading globally with new music genres (Drake, 2012), to describe various bleak and extractive urban enclosures and assemblages. As City- and select partner-run 'trap' parks, plazas, and streetscapes, recent public improvement efforts in downtown San Antonio appear to have ensnared at least some of their desired targets, as the City's Downtown Food Truck Program continues into a second year of operation and thousands of citations have been issued and numerous arrests made in recent months for violations of local ordinances meant to discourage the homeless or working poor from asking for assistance or simply being present downtown (Garcia, 2015). Beyond mainly confronting, detaining, and, in some cases, temporarily ejecting visibly poor or frugal residents and visitors from the downtown area, these policing efforts appear to help regularize frequent police interjections as well as economic segregation in downtown public places including parks.

Current efforts to deter less affluent or more frugal residents and visitors from downtown San Antonio appear to connect with US 'skyboxification' (Morello and Melinik, 2013), or the isolation of the wealthy in hyper-restricted residential developments as well as new perceivable 'status' enclosures or offerings (Atlas, 2013) in publicly accessible places. From new fee-based expedited security screening lanes at some US airports including as trialed in San Antonio, where priority lanes run parallel with and observable to more congested non-fee lanes, to *diner en blanc* events in public sites and parks, such practices decrease opportunities for social mixing across widening economic divides as well as help to normalize new or resurgent forms of economically tiered access to a wide range of publicly supported places and public services.

Just as San Antonio has been identified as a bellwether for how the state of Texas and the nation will likely grow in terms of ethnicity (Dotan, 2012), the city also appears to be marking new territory in a movement towards or return to extreme economic inequality and segregation in the US, as influenced by historical and current patterns of urban marginalization; these patterns include in San Antonio unique Spanish colonial era 'frontier' socio-spatial logics (Teja, 1996, pp. 22–29) as well as historically more recent and 'American' racial and ethnic segregation shaping the city (Drennon, 2012), along with class- and cadre-based distinctions informing migration within and from Mexico to specific sections of San Antonio in different eras (Cave, 2014). Although more prevalent views of US urban gentrification, marginalization, or 'revanchism' (Smith, 1996) tend to assume 'Anglo' domination, or 'swarms of . . . white people moving into . . . minority neighborhoods' (Hartnett, 2014) and 'normatively white' and 'middle-class' place-making approaches (Elwood, Lawson, and Novak, 2015, p. 126), current practices and patterns in San Antonio might complicate such understandings. At least, they might help demonstrate how formal place-making approaches incorporating key elements of what has been described as the US *barrio* 'model of urbanism' (Diaz and Torres, 2012, p. 22), and specifically mobile food truck and trailer vending, might not necessarily lead to more equitable urban planning outcomes or better conditions for lower-income urban residents and visitors, including in US cities with majority, economically diverse, and growing *Latin@* populations.

'Pop-up' dog bakeries: making space for *zombis fresas*

On a recent weekend afternoon, I caught the tail end of a more dog-centric Urban Pet Market now held regularly at Travis Park. The new market is apparently one of a number of 'new shopping experiences' (Arias, 2014) and special events that, along with City-selected and closely monitored food truck vending and other interventions, is meant to help reform or 'redirect' the downtown San Antonio area (Chasnoff, 2014). A desired transition that has been framed by some as moving San Antonio away from a being a city of 'two downtowns', one wealthy/vital and one poor/'ghost', to just one (Rivard, 2014).

Earlier in the year, an attempt by a church that neighbours Travis Park to reserve the park for an annual outdoor gathering which engages hundreds of volunteers and church members in providing goods and services for 'needy individuals and families', including fellow congregants, was denied until a City Council member intervened on behalf of the event organizer (Chasnoff, 2014). City staff members managing the booking of the revitalized Travis Park, which can be hired for various events including private weddings, apparently informed the church representative initially that the 'repurposed' Travis Park could not accommodate the gathering in its 18th consecutive year or future years, as 'the homeless need to be out of the downtown area' (quoted in Chasnoff, 2014).

With an estimated 150,000 feral dogs and more than 2,000 reported dog bites in San Antonio in 2013 (Davis, 2013), it is not surprising that City staff members managing Travis Park might perceive animal welfare as a public concern worthy of the regular transformation of the municipal park into a commercial 'pet market' to, in part, generate funds to help support local animal services; on the day I attended the market at Travis Park, it included a pop-up bakery for dogs, a truck-based mobile pet grooming business, and other temporary services for pet owners with disposable income. Although I am a dog owner and comfortable around dogs, I was somewhat taken aback to find this day within Travis Park not only mobile vending catering to pets but two large and potentially aggressive dogs roaming unleashed in the park (see Figure 4.4), within sight of their owner as well as a uniformed police officer and not far from a busy street and public bus stop. This is especially notable, simply considering the astounding number of San Antonio *ciudand@s* to have suffered or witnessed dog attacks, but also remembering the

Figure 4.4 Large dogs enjoying an 'unleashed' moment outside of the designated 'dog park' enclosure at Travis Park during a recent Urban Pet Market event.

Source: Mark Tirpak, 2014.

new purpose-built off-leash 'dog park' enclosure and posted rules at Travis Park, where glass soda bottles have apparently been deemed a serious public safety risk and quality-of-life concern worthy of police action.

Heading towards the posted rules at Travis Park, I spotted no barriers to utilizing the empty dog enclosure as the two large and unleashed dogs enjoyed other sections of the park unmolested. I did, however, notice being 'tailed', or followed at close distance, by the observed police officer as I explored the empty dog enclosure and read the posted park rules. Although the officer did not engage with me directly, his *de facto* 'stop-and-frisk' (Goldstein, 2014) made it clear that, once again, it was time for me to exit Travis Park.

As I left Travis Park for the day, I noticed a group of people waiting for public buses, their backs to the closing Urban Pet Market and unleashed dogs. With no electronic signs to alert them to arrivals or delays, some waiting passengers stood at the edge of the sidewalk as well as in the traffic lane scanning for signs of the next buses. The human need for a toilet this particular day could have easily cost one of these waiting passengers an unwanted encounter with a police officer or unleashed dog while crossing the park in search of a public bathroom. Failing to find one and depending on their appearances, a bus passenger would likely have to endure additional surveillance and possibly experience being barred from using the toilet facilities within the lobby of a nearby hotel, despite this avenue sometimes being suggested to park-goers. Alternatively, using the bathroom of a neighbouring restaurant would likely require a small purchase or tip for a staff member. This reinforces how, in many regards, Travis Park and the surrounding downtown district have not been remade or managed with the health or quality of life of many San Antonio residents and visitors in mind.

Conclusion

With this chapter, I have attempted to explore emerging formal place-(re)making assemblages in downtown San Antonio, Texas, that appear to marginalize a wide body of residents and visitors, including by directing limited public resources towards attempting to serve the strange whims of imagined downtown visitants. Wealthy, mindless visitors, or *zombis fresas*, who presumably seek $12USD street taco lunches and are without need for public toilets or hand washing facilities. Or who want to spend time in an outdoor cage enjoying baked goods with pets. Or who long to enjoy expensive picnic dinners clad in white under the shadow of a Confederate memorial, as catered by workers earning less than a living wage. While the police are dispatched to remove park benches, levy fines, and otherwise disrupt public opportunities and offerings for the city's poorest or most frugal. A grounded and more 'auto' ethnographic perspective, attentive to the socio-materiality of urban space, allows these unusual public expenditures and trends, which potentially mark the leading edges and convergences of economic segregation and demographic and cultural changes in the US, to be brought to light. This not only suggests that aspects of *Latin@* urbanism and transforming poverty politics in the US invite further investigation, but, ideally, prompts new action taking in San Antonio to challenge planned or publicly supported economic marginalization in the city.

Note

1 Quotes from individuals presented in this chapter are reconstructed from memory of the encounter.

References

Amin A., 2013. Telescopic urbanism and the poor. *City*, 17(4), pp. 476–492.

Andreou, A., 2015. Anti-homeless spikes: 'sleeping rough opened my eyes to the city's barbed cruelty'. *The Guardian* [online] 18 February. Available at: <http://www.theguardian.com/society/2015/feb/18/defensive-architecture-keeps-poverty-undeen-and-makes-us-more-hostile>.015.

Arellano, G., 2005. !Ask a Mexican! *OC Weekly* [online] 11 November. Available at: <http://www.ocweekly.com/2005-11-10/columns/ask-a-mexican/>.015.

Arellano, G., 2012. *Taco USA: how Mexican food conquered America*. New York: Schribner.

Arias, S., 2014. Downtown pet market helps no-kill cause. *San Antonio Express-News* [online] 19 July. Available at: <http://www.mysanantonio.com/news/local/article/Downtown-pet-market-helps-no-kill-cause-5633079.php>. 015.

Atlas, J., 2013. The great divide: class struggle in the sky. *The New York Times* [online] 14 July. Available at: <http://opinionator.blogs.nytimes.com/2013/07/06/class-struggle-in-the-sky/>. 015.

Bailey, W., 2014. Southwest Airlines to help fund makeover of historic Travis Park. *San Antonio Business Journal* [online] 26 March. Available at: <http://www.bizjournals.com/sanantonio/blog/2014/03/southwest-airlines-to-help-fund-makeover-of.html>. 015.

Blaze, M., 2014. 10 Día de los Muertos events to attend this weekend. *San Antonio Current*, [online] 30 October. Available at: <http://www.sacurrent.com/Blogs/archives/2014/10/30/10-dia-de-los-muertos-events-to-attend-this-weekend>. 015.

Burgen, S., 2015. US now has more Spanish speakers than Spain – only Mexico has more. *The Guardian* [online] 29 June. Available at: <http://www.theguardian.com/us-news/2015/jun/29/us-second-biggest-spanish-speaking-country>. 015.

Carman, T., 2009. Tacos al pastor: Lebanon's gift to Mexico. *Washington City Paper* [online] 12 November. Available at: <http://www.washingtoncitypaper.com/blogs/youngandhungry/2009/11/12/tacos-al-pastor-lebanons-gift-to-mexico/>. 015.

Casura, L., 2015. The big sort: San Antonio scores high on economic segregation. *The Rivard Report* [online] 9 March. Available at: <http://therivardreport.com/big-sort-san-antonio-scores-high-economic-segregation/>. 015.

Cave, D., 2014. The way north, day 5: The 'ambiente' of equality. *The New York Times* [online]. Available at: <http://www.nytimes.com/interactive/2014/us/the-way-north.html?_r=0#p/5>. 015.

Chasnoff, B., 2014. Bernal saves downtown event for homeless. *San Antonio Express-News* [online] 26 September. Available at: <http://www.expressnews.com/news/news_columnists/brian_chasnoff/article/Bernal-saves-downtown-event-for-homeless-5783648.php>. 015.

City of San Antonio, n.d. *Parks Safety & Rules* [online]. Available at: <http://www.sanantonio.gov/parksandrec/rules.aspx>. 015.

City of San Antonio. 2014. *Rules and regulations for Downtown Food Truck Program*, [online] 29 January. Available at: <http://www.sanantonio.gov/Portals/0/Files/dtops/DTMobileFoodTruckRulesandRegulations.pdf> [Accessed xxxx 2015].

Davila, V., 2015. Forum takes on gentrification in San Antonio. *San Antonio Express-News* [online] 11 February. Available at: <http://www.expressnews.com/news/local/article/Gentrification-Town-Hall-6074177.php>. 015.

Davis, V.T., 2013. City's massive stray problem despite no-kill goal success. *San Antonio Express-News* [online] 20 July 2013. Available at: <http://www.expressnews.com/news/local/article/City-s-massive-stray-problem-despite-no-kill-goal-4676873.php#/0>. 015.

Demby, G., 2013. 'Latin@' offers a gender-neutral choice; but how to pronounce it? *National Public Radio* [online] 7 January. Available at: <http://www.npr.org/blogs/thetwo-way/2013/01/07/168818064/latin-offers-a-gender-neutral-choice-but-how-to-pronounce-it>. 015.

Diaz, D.R. and Torres, R.D., eds., 2012. *Latino urbanism: the politics of planning, policy, and redevelopment*. New York: New York University Press.

Dotan, T., 2012. Latino Texas: 'San Antonio is a peek at the America of tomorrow'. *The Guardian* [online] 6 October. Available at: <http://www.theguardian.com/world/2012/oct/06/san-antonio-texas-elections-democrats-republicans>. 015.

Drake, D., 2012. Real trap sh*t? The commodification of southern rap's drug-fueled subgenre. *Complex* [online] 9 October. Available at: <http://www.complex.com/music/2012/10/real-trap-sht-the-commodification-of-southern-raps-drug-fueled-sub-genre>/. 015.

Drennon, C., 2012. The roots of economic segregation. *San Antonio Express-News* [online]. Available at: <http://www.mysanantonio.com/opinion/commentary/article/Restrictive-policies-economically-segregated-3779323.php>. 015.

El Daily Post, 2015. Income disparity in Mexico is dizzying. . .and growing [online]. Available at: <http://www.eldailypost.com/news/2015/06/income-disparity-in-mexico-is-dizzying-and-growing/>. 015.

Elwood, L., Lawson, V., and Nowak, S., 2015. Middle-class poverty politics: making place, making people. *Annals of the Association of American Geographers,* 105(1), pp. 123–143.

Everett, J., 2012. Perceptions of poverty. *Huffington Post* [online] 17 September. Available at:<http://www.huffingtonpost.com/jeremy-everett/working-poverty-_b_1680831.html>. 015.

Gandy, M., 2005. Cyborg urbanization: complexity and monstrosity in the contemporary city. *International Journal of Urban and Regional Research*, 29(1), pp. 26–49.

Garcia, G., 2015. City's homeless crackdown reaches a new low. *San Antonio Express-News* [online]. Available at: <http://www.expressnews.com/news/news_columnists/gilbert_garcia/article/City-s-homeless-crackdown-reaches-a-new-low-6193914.php>. 015.

Gold, J., 2012. How America became a food truck nation. *Smithsonian Magazine* [online]. March. Available at: <http://www.smithsonianmag.com/travel/how-america-became-a-food-truck-nation-99979799/>. 015.

Goldstein, J., 2014. Safer era tests wisdom of 'broken windows' focus on minor crime. *The New York Times* [online] 24 July. Available at: <http://www.nytimes.com/2014/07/25/nyregion/safer-era-tests-wisdom-of-broken-windows-focus-on-minor-crime-in-new-york-city.html?smid=tw-share>. 015.

Gonzalez, J., 2014. Bexar County considers minimum wage increase. *San Antonio Express-News* [online]. Available at: <http://www.mysanantonio.com/news/local/article/Bexar-County-considers-minimum-wage-increase-5901556.php>. 015.

Greer, G., 2015. White out rain check: dinner en blanc to transform Travis Park. *The Rivard Report* [online] 5 March. Available at: <http://therivardreport.com/white-dinner-en-blanc-transform-travis-park/>. 015.

Hartnett, K., 2014. Gentrification: white people following white people. *The Boston Globe* [online] 6 August. Available at: <http://www.bostonglobe.com/ideas/2014/08/06/gentrification-harvard-study-using-google-street-view-shows-that-white-people-follow-white-people/Pv29PzqMtp8Kc69ctpVc5J/story.html>. 015.

Henderson, A., 2015. 10 cities where an appalling number of Americans don't have enough food. *Alternet* [online] 7 January. Available at: <http://www.alternet.org/news-amp-politics/10-cities-where-appalling-number-americans-dont-have-enough-food>. 015.

Johnson, T., 2013. In Mexico, 'juniors' and 'ladies' find scorn in social networks. *McClatchy Washington Bureau* [online] 10 May. Available at: <http://www.mcclatchydc.com/2013/05/10/190952/in-mexico-juniors-and-ladies-find.html>. 015.

Lauro, S.J. and Embry, K., 2008. A zombie manifesto: the nonhuman condition in the era of advanced capitalism. *boundary 2,* 35(1) pp. 85–108.

McInnis, J., 2014. Call for entries: twisted Taco truck throwdown. *San Antonio Express-News*, [online] 11 March. Available at: <http://blog.mysanantonio.com/food/2014/03/call-for-entries-twisted-taco-truck-throwdown/>. 015.

Miller, L., 2015. Heroin is a white-people problem: bad medicine, economic rot and the enterprising Mexican town that turned the Heartland on to black tar. *Salon* [online] 18 April. Available at: <http://www.salon.com/2015/04/18/heroin_is_a_white_people_problem_bad_medicine_economic_rot_and_the_enterprising_mexican_town_that_turned_the_heartland_on_to_black_tar/>. 015.

Morello, C., and Melinik, T., 2013. Washington: A world apart. *The Washington Post* [online] 9 November. Available at: <http://www.washingtonpost.com/sf/local/2013/11/09/washington-a-world-apart/>. 015.

Olivo, B., 2013. Updated: public improvement district likely to expand. *San Antonio Express-News* [online] 21 May. Available at: <http://blog.mysanantonio.com/downtown/2013/05/centro-to-tap-into-public-improvement-district/>. 015.

Parker, K., 2014. Revitalized Travis Park reopens. *San Antonio Express-News* [online]. Available at: <http://www.expressnews.com/news/local/article/Revitalized-Travis-Park-reopens-5365509.php#/5>. 015.

Pilcher, J.M., 2012. *Planet taco: a global history of Mexican food*. Oxford: Oxford University Press.

Rivard, R., 2013. People want a park: San Antonio's passion for Hemisfair. *The Rivard Report* [online] 28 May. Available at: <http://therivardreport.com/people-want-a-park-san-antonios-passion-for-hemisfair/>. 015.

Rivard, R., 2014. Ghost buildings haunt San Antonio's 'Decade of Downtown'. *The Rivard Report* [online] 24 February. Available at: <http://therivardreport.com/ghost-buildings-san-antonios-decade-of-downtown/>. 015].

Robert Woods Johnson Foundation, 2012. ¡Por Vida!: creating healthy menu options in San Antonio. *Healthy Kids, Healthy Communities: Supporting Community Action to Prevent Childhood Obesity* [online] April. Available at: <http://www.healthykidshealthycommunities.org/%C2%A1por-vida-creating-healthy-menu-options-san-antonio-0>. 015.

Rueda, M., 2014. Stuffed chili spurs conversation about classism in Mexico. *Fusion* [online] 28 August. Available at: <http://fusion.net/story/6395/stuffed-chili-spurs-conversation-about-classism-in-mexico/>. 015.

Sanders, H., 2014. Cityscrapes: lands of the lost. *San Antonio Current* [online] 17 December. Available at: <http://www.sacurrent.com/sanantonio/cityscrapes-lands-of-the-lost/Content?oid=2359149>. 015.

Sanders, H., 2015. Hotel junkie: do big boom projects help average people In SA? *San Antonio Current* [online] 4 March. Available at: <http://www.sacurrent.com/sanantonio/hotel-junkie-do-big-boom-projects-help-average-people-in-sa/Content?oid=2404294>. 015.

Savičić, G. and Savić, S., eds, 2012. *Unpleasant design*. Belgrade: G.L.O.R.I.A.

Smith, Neil. 1996. *The new urban frontier: gentrification and the revanchist city*. London: Routledge.

Spener, D., 2010. Movidas rascuaches: strategies of migrant resistance on the Mexico-U.S. border. *Aztlán: A Journal of Chicano Studies*, 35(2), pp. 9–36.

Tedesco, J., 2014. Police arrest homeless woman taken in by Starbucks. *San Antonio Express-News* [online], 24 October. Available at: <http://blog.mysanantonio.com/downtown/2014/10/police-arrest-homeless-woman-taken-in-by-starbucks/#27855101=0>. 015.

Teja, J.F.D.L., 1996. *San Antonio de Béxar: a community on New Spain's northern frontier*. Albuquerque: University of New Mexico Press.

Villa, M., 2012. Spanglish as a third language. *San Antonio Express-News* [online], 19 September. Available at: <http://www.mysanantonio.com/sacultura/conexion/article/Spanglish-as-a-third-language-3877567.php>. 015.

Wong, V., 2013. The Mexican Coca-Cola myth: it's almost American. *Bloomberg Businessweek*, [online] 11 November. Available at: <http://www.businessweek.com/articles/2013–11–11/the-mexican-coca-cola-myth-its-almost-american>. 015.

Ybarra-Frausto, T., 1991. Rasquachismo: a Chicano sensibility. In Griswold del Castillo, R., McKenna, T. and Yarbro-Bejarano, Y., eds., 1991. *Chicano art: resistance and affirmation, 1965–1985*. Los Angeles: Wight Art Gallery, University of California, Los Angeles. pp. 155–162.

5 Cities that are just cities

AbdouMaliq Simone

Delimiting justice

While marginality has long been posited as a condition of radically uneven distributions of opportunities and resources within urban life and as an instrument of urban structuration itself, to what extent are contemporary manifestations of the urban marginal to the sustenance of urban life in general? The de-skilling of urban populations, their enforced circumscription in specific socio-spatial arrangements, and the continuous wounding of bodies and their efforts that force large numbers into incessant compensations perhaps renders ecological footprints and environmental forces more salient than current preoccupations with the anthropocene.

Yet, if rendering marginal the possibilities of life without substantial debilitating constraints seems impossible to conceive as a plausible urban disposition, how is it possible to have confidence in the by-now common conclusion that the sustenance of human existence will require a redistribution of resources one way or another? In a long history of injustice, the likely prospect is simply that many residents will be removed to some form of remoteness or irrelevance, or that the city fulfils the dystopian fantasies of either total lockdown or everyday war (Graham, 2010; Sassen, 2010; Gregory, 2011).

A critical concern of a more broadly drawn notion of 'sustainability' is how to interrelate new ways of working, producing, residing, exchanging, and affecting to the practicalities of administering material flows, social circulations, and the distribution of livelihood opportunities (Stiegler, 1998, 2009; Hodson and Marvin, 2009; Lawhon and Murphy, 2011; Jaglin, 2014). Can a notion where the city is 'marginal' to any conceptual notions applied to it be a useful starting perspective through which to discover how to counter the ways in which cities marginalize many of their residents?

People with skills and ideas are crossing each other's paths all of the time. But only a small fraction of the capacities, actions, and value inherent in the possibilities of their more substantive interaction are ever considered or mined. What already might be taking place as inherent aspects of urban life are continuously marginalized. This includes the provision of various forms of care, the pooling of resources, the differentiated uses of shared objects and spaces, and the creation of new services (Dikec, Clark, and Barnett, 2009; DeVerteuil and Wilton, 2009;

Hart et al., 2010; Stewart, 2011). How do we do justice to these complexions of everyday life? Urban politics and governance could always do more, but instead there is usually a tendency to settle for less than what might be possible. But at the same time, justice is not a specific disposition or destination but a matter of timing, of something always being worked on, worked out (Soja, 2010; Fincher and Iveson, 2012; MacLeod and McFarlane, 2014).

This chapter is concerned with notions of justice but from a difficult angle. For instead of focusing on 'the just city', I will primarily consider the implications of the notion, 'just the city'. Here, the notion of *just* is immediately altered to connote, 'well, it's only the city, after all, what do you expect', as if certain aspirations and actualities somehow remain permanently out of reach. The city always resists a particular kind of specification that makes proliferating injustices visible *and*, at the same time, highlights the incalculable ways in which relations between inhabitants are continuously repaired and remade (Merrifield, 2012). While this can be a matter of everyday conviviality, what is more important is the continuous deployment and renovation of technique and practices that recalibrate and redistribute opportunities, burdens, and responsibilities across the intersections of contact that inhabitants must continuously navigate (Telles and Hirata, 2007; Nijman, 2010; Amin, 2013; King and Dovey, 2013; Dovey, 2014; Vasudevan, 2014).

The majority of city inhabitants in the so-called global South lived through highly entangled relationships with others of different backgrounds and capacities. They consolidated territorial and social distinctions while, at the same time incorporating each other into differentiated understandings about what it would take to survive the city. The commonality of these different understandings was that any practice, idea, or agenda of inhabitation was never going to be sufficient – there would always be more to do (Robinson, 2002; 2011; Edensor and Jayne, 2012). On the other hand, great value was placed on the notion of the incremental – to do things a little bit at a time, and each thing had its own time. Even if one could build all at once, or buy someone out completely, such moves were attenuated by the sense that 'time had its own time'. Wherever one was at the moment, it was good enough, for now – not in general, but for now. What was thus more important was the sense of cities as projects that were incomplete. As incomplete, then, nothing was foreclosed, wrapped, or summed up (Perlman, 2010; Chattopadhyay, 2012; Harms, 2013; Silver, 2014).

As such, districts in cities such as Jakarta, Sao Paolo, Karachi, and Bangkok were full of people who considered themselves eligible to make demands, propositions, and critique. The viability of such cities has largely depended on situations where many different things can take place in the same place at the same time without people judged or implicated by this plurality – where there are many different things and ways of life to witness, but where there is no obligation to witness them. Here, the dense proximities of different statuses, backgrounds, orientations, and capacities do not obviate the persistence of the distinctions within such categories (Holston, 2008; McFarlane, 2008, 2011; Hasan et al., 2010; McQuarrie et al., 2013).

Much of the interaction among residents never went anywhere, often never accomplished anything. But instead of this apparent failure acting as a deterrent, it tended to keep the game going. The difficulties of everyday life in places like Sao Paolo, Jakarta, and Karachi feed an intense hunger for justice and equity. But there is also a wariness of pinning things down too much, of instituting policies where capacities and conditions are calculated and compared. There is often a preference for keeping things incomplete.

Everyday life may be full of antagonisms, misconstrued behaviours, evasions, tricks, and manipulations, but these are also the conditions that give inhabitants something to work with, something to try and put right, something that brings people together who otherwise would keep their distance, and thus a platform for the incessant rehearsal of different ways to 'work things out' (Benjamin, 2008; Sánchez, 2008; Sundaram, 2009; Nielsen, 2010; Millar, 2014). Indeed this can be labour intensive and a strain for residents already overwhelmed with trying to make ends meet. But it is in these rehearsals where residents often feel that new vistas are opened up, where at least they are exposed to worlds otherwise inaccessible, even though there are no guarantees that they can take advantage of them. After all, it is just the city.

The problem of multiplicities

What do we mean when we say that the city is *just* the city? The phrase would seem to suggest a limited set of expectations should be brought to bear on what the city is and can do. Likewise, a sense of tolerance and patience might be displayed in the face of something, which is, after all, just what it is, nothing else and nothing more. Residents and observers could ask why the city does not do *x*, *y*, or *z*, or why the city behaves in the way it does, and our response here would then seem to indicate, while even dismissing such queries, that the city is simply in its past and present modes of operation doing what it was meant to do, or the only thing it can do, so that it is not reasonable to ask of it anything more than this.

Why is this play of semantics important? Much has been asked of the city. In the face of queries, demands, and expectations, there is presently a great deal of despondency and even alarm about what cities have become and could become. At a very basic level, cities were the embodiments of urbanization in a conjunction of process and outcome. The specializations of production and labour markets, the densification of residency warranted by the exigencies of profitable manufacturing, and the agglomeration of strangers with no over-arching claims of belonging produced the form of the modern city, and this form, in turn, exerted a recursive impact on the continuous shaping of urbanization processes. City and urbanization thus existed in an isomorphic relationship, each mirroring the other.

Today, the city is no longer the exemplar or the culmination of urbanization. Rather, it exists in a plural field of multilayered patchworks, a component in an extensive regionalization of both coordinated and disjointed production, inhabitation, and governance. Given its prior status, the pluralization of urbanization thus implicitly renders the city as a kind of anachronism. The city is not a sign of the proliferation and complexity of global urban change but a sign that the trappings of modernity closely associated with it are nearly finished. The city becomes museum, or a place of ruins.

This does not mean that cities do not remain important subjects of discourse, the locus of the biopolitical elaboration of populations, or as the harbingers of sustainable density needed in order to effective modulate the impact of the anthropocene. Rather, they are simply not the first place to look when trying to understand the trajectories of urbanization. But what the phrase *the city is just the city* would seem to connote is that the de-centering of urbanization away from the city or the folding in of the city to new territorialization and topological deformations does not change matters all that much. For the city remains *just* the city, something that persists no matter how urbanization is manifested.

The point is that there is a 'rendition' of the city that does not depend upon any specific narrative of its ontogenesis; that how things got to be in and part of the city, while critically important for any timing of developments for the city, does not necessarily specify what those things do with and for each other once they are *there*, ensconced as components of city life. Rather than asking the city to comprehensively embody urbanization, to say the city is just the city is to request that it be left alone. This might seem to be an unreasonable request given that it implicitly ignores where the city has come from, and where it might go. But this is precisely the point that such considerations raise: that the city has always suggested what Foucault calls 'the problem of multiplicities'.

Knowledge about the city – its spaces, populations, and dynamics – was grounded largely in measures taken to allow some circulations of urban bodies and to disallow others. In face of the impact of infectious diseases, the impetus was to enclose, survey, and prevent. Attempts to shape the vectors of disease, cultivate normative attitudes towards bodily conduct, and the put activities in their proper place reiterated the reality of 'demonic' circulations. These circulations pass across boundaries that had separated the human and inhuman, the living and non-living, and the reasonable and unreasonable. They marked the limits of the integrity of the individual body, which would have to be buttressed with a sense of its subjectification. As Eugene Thacker (2012) reminds us, the threat was less the microbes and viruses that were the agents of infection than the conduits of transgressive circulation themselves – i.e., the ways in which things in the city could reach, affect, and move through each other in ways that could slip through the available tools of apprehension and monitoring.

Once all of the dangerous, 'demonic' circulations are excluded so as to make the individual, normative subject predominant, the city as a political space disappears in the strict sense of deciding what it is without the intervention of a rational third party. For these ungovernable circulations are what urban government must capture in order to operate (Heron, 2011).

To ask to leave the city alone, to not expect it to deliver the kinds of answers, models, and capacities asked of it, to separate it from the historical processes that produced its forms of existence is to delve into the unreasonableness at the heart of its organization. Whatever human effort was at work that attempted to solve particular problems of existence or to render human existence as opportunistic instruments for domination on the part of some over others, what remains in the city are the possibilities of criss-crossing circulations among materials of all kinds in various intersections. In these intersections, human residents may be the predominant actors as the agents most capable of redirecting and remaking what is present within the city. But, the city – being *just* the city, and not the human instrument of other things – always slips away from the grasp of its human inhabitants, and we can never be certain of exactly what will show up or in what form, or what will forever disappear (Smith, 2010; Colebrook, 2012; Matts and Tynan, 2012; Woodard, 2013).

For the question is, what is in the city that makes it just what it is? And we must ask this in a way that does not demand of it to be different from domains that are *not* the city even when those domains may contain the very same things. In other words, how can one consider the city without having to add up its componential parts, render them a particular, game-changing, identity-establishing addition, or get the city to speak its structure or its frame?

Cities are full of all kinds of stuff. Not only stuff of different compositions and natures, but stuff taking place in different times of composition and decomposition. While human inhabitants act on this stuff, the stuff also acts on them. The city was important to modernity, to the idea that human effort could progressively develop the conditions through which human capacities would be infinitely extended, because it situated things in conditions where the intensities and scope through which things could exert an effect on each other generated complex rhythms of generation and modulation. These rhythms then could be lived through as processes of self-surpassing, where the complexity unfolding all around him could be read as characteristics of man's own being (Lorimer, 2012; Braidotti, 2013; Hinchliffe and Lavau, 2013).

The city being *just* the city doesn't mean that the city can be anything at all, for there are clear constraints to what materials can do under specific conditions. There are particular parameters of what and how things can be acted upon. But these are contingent necessities. They need not be what they are; they need not take the form that they take, even though they cannot take any form whatsoever (Meillassoux, 2010). As Virno (2009) indicates, work relies upon an 'inarticulate power which is not reducible to a series of pre-set potential acts, acquires an extrinsic, or better, pragmatic aspect in the commodity labour-power' (p. 101).

But at the same time, the ability to make use of such disorientation in a world opposed to any concretization of a definitive environment requires the proliferation of protocols, standards, credit checks, passwords, surveillance cameras, sensors, algorithmic decisions, and tracking devices. These are all methods for registering instability, for 'moving' with movement, by way of efforts to anticipate and calculate probabilities, and to 'see in advance' (Crandall, 2010). If urban economies try to harness the capacity of the city to be *just* the city – a locus of incessant disorientation, intersection, and movement – they also 'fill' the city with various devices for recording, modulating, detecting, speculating, and anticipating which themselves constitute unknown exertions of force, inclination, reverberation, and intensity that cannot be fully calculated (Parisi, 2012). So to say that the city is *just* the city is to point to the constant process of trying to catch up with that which any iteration of *the city* has to tell. There is an inevitable falling behind, of never reaching a final destination, even though the city has been there all along; it has no secrets to hide, no revelations waiting for the right moment or development stage.

While the intensities exerted by things and bodies may generate attractions and repulsions, draw things near or far, there are no predetermined reasons why things or events should necessarily connect, be in a relationship with each other. Once something is put together, it may in retrospect be able to say, 'Oh, yeah, these things were really made for each other – they have this kind of character and that kind'. It is possible after the fact to explain how things are attuned to each other, complement or engender new capacities. But these characteristics or properties of the relationship cannot fully account for how the relationship was actually put together. One needs to see what makes these relationships specific, makes them hold or endure across different times and situations, while at the same time being able to become something else.

The city as exit

Throughout the global South there appears to be a seamless convergence between the desire to live in efficiently managed high-rise complexes away but still proximate to the urban core and the prolific provisioning of residential units, cut-rate mortgages, subsidized rentals, and continuously elaborated lifestyles. The refitting of the urban core with high-end, multinational services and amenities also targets residents of poor and working-class areas as objects of eviction and enticement. They are to be vacated to distant peripheries. Much of the violence of forced removals of past years has been replaced with promises of free televisions and washing machines to accompany relocation to cheaply built high-rise accommodation at the periphery of the urban region.

But the vacating of worlds has also been a feature of cities for a long time – something reflected in the convoluted patchworks of built environments full of materials emplaced in strange contiguities. Buildings of all kinds and functions in many parts of cities, particularly those mega-regions of the South, are entangled in seemingly improbable embraces, reflecting the bodily movements of a plethora of creatures – the circuitries of ants, spiders, rodents, snakes, as well as different forms of microbial life (see Swyngedouw, 2006; McFarlane, 2008; Martínez-Alier et al., 2009; Hagan, 2015 for a more general discussion).

I work in inner-city districts of Jakarta still full of an intense mixture of social backgrounds, economies, and built environments. While residents can recall particular sequences of inhabitation, construction, renovation, removal, and destruction in their immediate built environment, they are often hard pressed to account for how its current structuring generated the image it assumes as a whole.

The differences in the morphology of districts that began their existence as a result of systematic planning and those that have emerged through various times and logics of self-construction can sometimes be indiscernible. Those districts that were laid out with well-ordered grids and demarcations of property could mutate in different hybrid forms. The back plots of constructed pavilions could be rented to others who would assemble makeshift structures and, over time, these ancillary settlements could generate a life of their own, proliferating across contiguous back plots, generating further subdivisions or consolidations until all available acreage was filled. Original tenants could intervene in reshaping the arrangements at the anterior, just as those of the anterior might be able to, over time, 'make moves' on the frontage. On contiguous blocks, none of this might necessarily occur. The 'seeding' and spread of such mutation assumed no 'usual' trajectory. Not dissimilarly, the more random and dispersed insertions of residents in districts that were largely constructed from the ground up with few interventions from municipalities or developers could also tend towards more linear gridded layouts over time, particularly in situations where there existed large-scale collective sentiment to aim for regularization of tenancies.

The point is that the built environment can express the making of relations that operate according to intensely experimental inclinations, as residents attempt to both differentiate themselves from each other yet, at the same time, find multiple ways to coexist. The ongoing tensions between fitting into a larger collective schema that orients residents to specific forms of optimal functioning – about how to be visible in such a way as to gain access to services and opportunities – and the need to also 'stand out' so as to make specific claims and garner particular opportunities are incessant. These tensions are clearly marked in built environments, and particularly in situations in which inhabitation meant not only having access to shelter but also a physical platform from which individuals and households could conduct economic activity and communicate specific messages about who they were and what they were capable of doing to a larger surrounds.

Built environments reflect assessments about what can be afforded, which include the trade-offs amongst various facets of everyday experience: Levels of comfort, resource consumption, use of space, public visibility, ease of access, aesthetic preference, flexibility of adjustment, security of tenure, personal safety, and geographic location are all weighed in terms of each other. In the intensely mixed-income, mixed-social background districts of central Jakarta, residents would navigate across a wide range of decisions – both their own and those of others.

Some residents would commit to substantial initial expenditures, building big, in order to hedge against future land divisions and to use the substantial outlay as a way of consolidating particular local advantages, which might include establishing a local power base. Others adopted a more wait-and-see attitude, preferring to install incremental adjustments over time, flexibly moving with the changes of a district's position in relationship to the wider city. The results of many different decisions as to how to deploy available resources and space existed in close proximity to each other, even when they did not necessarily have anything directly to do with each other.

Experimentation was a process, however, incumbent in living in the midst of this proximity. As Wita, a 49-year-old market manager put it, 'We were not looking for mirrors, we wanted to jump beyond what we knew because we were constantly reminded that there was something really possible that we were not aware of, not just silly dreams, but real things that could happen if we only could only hitch our wagons to them'. When residents move day by day through an intense mixture of living situations marked by the extensive diversity in their field of vision how, as Ardhi, a 35-year-old artist muses, 'are we going to tell what is proper and what is not, what works or what is useless?'

Here, the multitudes of small initiatives that people made to both make all of the ends meet and to keep the mixtures from being too overwhelming made 'staying in place' itself an ordeal. Yohannes, a bus driver, points out, '[in our district] we knew each pretty well; we knew what to expect and everyone knew that if they got too much out of line they would get what was coming to them; people knew that they were in for big trouble if they were to steal and cheat or get violent, but what was always a little bit strange was the way in which people would do all of these little things that were just a little bit unusual, the way someone might stop someone in the middle of the street they didn't know and simply tell them a little something about something taking place somewhere else, or the strange way they might decorate their door, or the way they might walk the streets in the middle of the night looking for god only knows what, or the way they might invite total strangers to sit and drink coffee with them in the front of the house'.

But as Abaye, the 'unofficial' manager of a fruit and vegetable market pointed out, 'All these small things were simply a way for us to get up the courage and to get familiar with attempts to make something big happen, to look at the places where we worked and lived as something more than that, as full of hidden secrets and mysteries that could be turned into something useful and without it seeming that we were doing something big, because that would only get us into trouble, with our neighbors, the police and the authorities'.

Susanto, a local headman in the district of Tanah Sereal, indicated that 'residents here were never content to just leave things alone. I mean they left each other alone, didn't bother them too much, but most people were always trying to see what could happen if they just went a little bit out of their way to buy something unusual to sell on the street, or to hand out pamphlets about how to start businesses to people waiting by the bus stop, or to sit on the corner and make notes about everyone who passed by, or who counted every vehicle that passed on a particular street, or who went door to door offering to repair things, or to invent machines for cleaning the water in the local canals, or to tutor the kids in the neighbourhood about how to fix engines'. Residents in Tanah Sereal would constantly talk about how unpredictable their neighbours could be, not in a bad or dangerous way. Rather, neighbours would often be seen hard at work trying to build something that probably wouldn't work or last, talk to people they had no prior experience with, and demonstrate a keen interest in some particular aspect of neighbourhood life – the passing of birds, the collection of unusual refuse, the composition of soil and streets, the loading and unloading of vehicles, or the rhythms of children's games – that otherwise would pass unnoticed.

To do justice

In highly differentiated fields of aspiration, history, capacity, feeling, and sense, actualized rights to the city can only be concretized in constituting spaces of differentiated action. Inhabitants would have to approach each other with a certain indifference, where the playing out of scenarios of action do not necessarily embody a valuation of other scenarios, neither passing judgement nor demanding relevance or affiliation.

Yet judgements have to be made; inhabitants simply cannot do what they want. No matter how well the city is surveyed, its population made known, no matter of big the data is, these matters of judgement still are those of political decision. A decision must be made to render some exertions of rights illegitimate. But 'who' decides in a situation when sovereign decisions of the past have only been able to specify concretizations of togetherness based on division and separation? The divisions are not necessarily those of 'friend' and 'enemy', but of what counts and what will not count, or will only count with some kind of negative connotation. If, more than ever, the efficacy of the city depends upon the constellation of various forms of enactment on the part of the heterogeneous things that inhabit the city, what amongst them can decide, who accords the space for decision and the terms for its registration? As we become more familiar with the self-annihilating characteristics of human thought and self-regard, only the submission of such thought to decisions other than that which it renders would seem to enable the endurance of human as species (Colebrook, 2012; Wolfe, 2013; Connolly, 2013). But what specifically do we pay attention to in this regard?

If the impetus for an urban modernism that has gotten us into so much trouble today aimed for the possibilities of continuous transformations, and as those transformations in human capacity with all of its concomitant prosthetics and augmentations parallel in speed and scope the human impact on the earth, to what extent is the just city predicated on a notion of the *just-human*? Not a human that is just, but an inhabitant who exemplifies its contradictory mélange of virtuous and destructive forces, an inhabitant who perhaps does more by doing little, and where the trajectories of striving mitigate an equally significant desire to stay still, to wait (Cohen, 2012).

Throughout six years of research in Jakarta asking residents how their neighbourhoods got to be the way there were, I would hear over and over again the invocation of a wily, sometimes even disingenuous ignorance, 'we cannot be so sure about what happened'. Obviously such a response contains specific ideas, but as soon as they are posited, as soon as residents hear themselves bringing voice to them, these ideas immediately sound either foolish or something to be withdrawn, as if their impact, no matter how correct these propositions might be in their analytical capacity, would inevitably sound false.

These acts of standing down, of hesitating, were not the result of the informant's distrust in being asked, for there was often an eagerness to talk about this, something not so usual for Jakartans long accustomed to reticence bordering on hostility. Nor did it seem this act of withdrawing came from any sense of humility. Rather, as Lia, a 37-year-old teacher puts it,

> to have a clear idea about something means you have to have a clear course of action. Just because the people in my neighbourhood were all pretty much

> in the same situation, facing the same daily realities, didn't mean that they all saw things the same way; in fact quite the opposite, but you wouldn't really know this for sure; it was just something you felt, because if people really expressed how differently they saw things, it would mean that there would be some kind of related action on the way, maybe not now, maybe not ever, but you would assume it, and so people simply let each other be. Of course around here no one could keep quiet for more than two seconds so one way or another, those points of view were always going around. A lot of times you wouldn't know where some specific idea or opinion was coming from, as people might say that they heard this from somewhere else, even though you probably knew that it was their idea anyway. It might seem as if we were hesitant to act decisively, but I think it was more a matter of letting a lot of things happen on their own.

There is a substantial history in cities like Sao Paolo, Mexico City, Karachi, and Jakarta of neighbourhoods trying to work themselves in and out of an array of problems and opportunities that are usually thoroughly intermeshed. These are places where opportunity becomes problem, problem becomes opportunity, and so forth. As the informants above indicate, how these things happened – how neighbourhoods would run into trouble, or sort things out, or continuously revise themselves or adamantly remain almost the same – probably entailed so many different kinds of explanation and factors that the sheer act of explaining could be construed as arrogant or even dangerous. This didn't mean that things were left inexplicable. Rather, the inexplicable was the occasion for further attempts to adjust or simply to wait. These behaviours and practices then seem to constitute an important archive for how to live urban life now, of how to create different kinds of stories about what is taking place even though the only evidence we have to go on is simply that the stuff in these stories feels possible. Marginalizing them cannot simply be the automatic response of the absence of validation.

To do justice, then, to promote a more just city, is to think of concrete ways of revitalizing the faith that residents once had in each other's ability to look closely at what each other did but at the same time do so without debilitating judgement, and to find ways to fold them into one's life but also to leave them alone to come up with things that would make folding something one would do on a continuous basis. For if everyone did the same thing, there would be nothing to fold, nothing to challenge, nothing to pay attention to or leave alone.

References

Amin, A., 2013. Surviving the turbulent future. *Environment and Planning D: Society and Space*, 31, pp. 140–156.

Benjamin, S., 2008. Occupancy urbanism: radicalizing politics and economy beyond policy and programs. *International Journal of Urban and Regional Research*, 32, pp. 719–729.

Braidotti, R., 2013. *The posthuman*. Cambridge: Polity.

Chattopadhyay S., 2012. *Unlearning the city: infrastructure in a new optical field.* Minneapolis, MN; London: University of Minnesota Press.

Cohen, T., 2012. Anecographics climate change and 'Late' deconstruction. In H. Sussman ed., *Impasses of the post-global volume 2,* Ann Arbor: Open Humanities Press, pp. 32–57.

Colebrook, C., 2012. A globe of one's own: in praise of the flat earth. *SubStance,* 41(127), pp. 30–39.

Connolly, W.E., 2013. The new materialism and the fragility of things. *Millennium: Journal of International Studies,* 41, pp. 399–412.

Crandall, J., 2010. The geospatialization of calculative operations: tracking, sensing, and megacities. *Theory Culture Society,* 27, pp. 68–90.

DeVerteuil, G. and Wilton, R., 2009. Spaces of abeyance, care and survival: the addiction treatment system as a site of regulatory richness. *Political Geography,* 28, pp. 463–472.

Dikec, M., Clark, N. and Barnett, C., 2009. Extending hospitality: giving space, taking time. *Paragraph,* 32, pp. 1–14.

Dovey, K., 2014. Incremental urbanism: the emergence of informal settlements. In Haas, T. and Olsson, K. eds., *Emergent urbanism: urban planning and design in a time of structural and systemic change.* Surrey, UK; Burlington, VT: Ashgate, pp. 45–53.

Edensor, T. and Jayne, M., eds., 2012. *Urban theory beyond the West: a world of cities.* London; New York: Routledge.

Fincher, R. and Iveson, K., 2012. Justice and injustice in the city. *Geographical Research,* 50, pp. 231–241.

Graham S., 2010. *Cities under siege: the new military urbanism.* London: Verso.

Gregory D., 2011. The everywhere war. *The Geographical Journal,* 177, pp. 238–250.

Hagan, S., 2015. *Ecological urbanism: the nature of the city.* London; New York: Routledge.

Harms E., 2013. Eviction time in the New Saigon: temporalities of displacement in the rubble of development. *Cultural Anthropology,* 28, pp. 344–368.

Hart, K., Laville, J. and Cattani, A.D. eds., 2010. *The human economy: a citizen's guide.* Cambridge: Polity.

Hasan, A., Sadiq, A. and Ahmed, S., 2010. Planning for high density in low-income settlements: four case studies from Karachi. *Urbanization and Emerging Population,* Issues-3, Human Settlements Working Paper Series. IIED and UNFPA.

Heron, N., 2011. The ungovernable. *Angelaki,* 16, pp. 159–174.

Hinchliffe, S. and Lavau, S., 2013. Naturecultures? Science, affect and the non-human. *Environment and Planning D: Society and Space,* 31, pp. 259–274.

Hodson, M. and Marvin, S., 2009. Cities mediating technological transitions: undertaking visions, intermediation and consequences. *Technology Analysis & Strategic Management,* 21, pp. 515–534.

Holston, J., 2008. *Insurgent citizenship: Disjunctions of democracy and modernity in Brazil.* Princeton, NJ: Princeton University Press.

Jaglin, S., 2014. Regulating service delivery in Southern cities: re-thinking urban heterogeneity. In S. Parnell and S. Oldfied, eds., *The Routledge handbook on cities of the global south,* pp. 434–448.

King, R. and Dovey, K., 2013. Interstitial metamorphoses: informal urbanism and the tourist gaze. *Environment and Planning D: Society and Space,* 31, pp. 1022–1040.

Lawhone, M. and Murphy, J., 2011. Socio-technical regimes and sustainability transitions: insights from political ecology. *Progress in Human Geography,* 36, pp. 1–25.

Lorimer, J., 2012. Multinatural geographies for the anthropocene. *Progress in Human Geography,* 36, pp. 593–612.

MacLeod, G. and McFarlane, C., 2014. Introduction: grammars of urban injustice. *Antipode,* 46, pp. 857–873.

Martínez-Alier, J., Kallis, G., Veuthey, S., Walter, M. and Temper, L., 2009. Social metabolism, ecological distribution conflicts and the language of valuation. *Capital, Nature, Socialism,* 20, pp. 58–87.

Matts, T. and Tynan, A., 2012. Geotrauma and the eco-clinic: nature, violence and ideology. *Symploke,* 20, pp. 154–171.

McFarlane, C., 2008. Urban shadows: materiality, the 'Southern City' and urban theory. *Geography Compass,* 2, pp. 340–358.

McFarlane, C., 2011. *Learning the city: knowledge and translocal assemblage.* Oxford: Wiley-Blackwell.

McQuarrie, M., Fernandes, N. and Shepard, C., 2013. The field of struggle, the office, and the flat: protest and aspirationin a Mumbai slum. *Public Culture,* 25, pp. 315–348.

Meillassoux, Q., 2010. *After finitude: an essay on the necessity of contingency.* New York: Continuum.

Merrifield, A., 2012. The politics of the encounter and the urbanization of the world. *City,* 16, pp. 269–283.

Millar, K., 2014. The precarious present: wageless labor and disrupted life in Rio de Janeiro, Brazil. *Cultural Anthropology,* 29, pp. 32–53.

Nielsen, M., 2010. Contrapuntal cosmopolitanism: distantiation as social relatedness among house-builders in Maputo, Mozambique. *Social Anthropology/Anthropologie Sociale*, 18, pp. 396–402.

Nijman, H., 2010. A study of space in Mumbai's slums. *Tijdschrift voor Economische en Sociale Geografie*, 101, pp. 4–17.

Parisi, L., 2012. Digital design and topological control. *Theory, Culture & Society*, 29, pp. 165–192.

Perlman, J.E., 2010. *Favela: four decades of living on the edge in Rio de Janeiro.* Oxford: Oxford University Press.

Robinson J., 2002. Global and world cities: a view from off the map. *International Journal of Urban and Regional Research,* 26, pp. 531–554.

Robinson, J., 2011. Cities in a world of cities: the comparative gesture. *International Journal of Urban and Regional Research,* 35, pp. 1–23.

Sánchez, R., 2008. Seized by the spirit: the mystical foundation of squatting among Pentacostals in Caracas (Venezuela) today. *Public Culture*, 20, pp. 267–305.

Sassen, S., 2010. When the city itself becomes a technology of war. *Theory, Culture and Society,* 27, pp. 33–50.

Silver, J., 2014. Incremental infrastructures: material improvisation and social collaboration across post-colonial Accra. *Urban Geography,* 35, pp. 788–804.

Smith, A.P., 2010. Believing in this world for the making of Gods: ecology of the virtual and the actual. *SubStance*, 39(121), pp. 103–114.

Soja, E., 2010. *Seeking spatial justice*. Minneapolis: University of Minnesota.

Stewart, K., 2011. Atmospheric attunements. *Environment and Planning D: Society and Space,* 29, pp. 445–453.

Stiegler, B., 1998. *Technics and time, 1: the fault of Epimetheus*. Stanford: Stanford University Press.

Stiegler, B., 2009. *Technics and time, 2: disorientation.* Stanford: Stanford University Press.

Sundaram, R., 2009. *Pirate modernity: media urbanism in Delhi.* London: Routledge.

Swyngedouw, E., 2006. Circulations and metabolisms: (hybrid) natures and (Cyborg) cities. *Science as Culture*, 15, pp. 105–121.

Telles, V.S. and Hirata, D.V., 2007. The city and urban practices: in the uncertain frontiers between the illegal, the informal, and the illicit. *Estudos Avançados,* 21, pp. 173–191.

Thacker, E., 2012. Notes on extinction and existence. *Configurations*, 20, pp. 137–148.
Vasudevan, A., 2014. The makeshift city: towards a global geography of squatting. *Progress in Human Geography*, Published online 30 April, 2014 DOI: 10.1177/0309132514531471.
Virno, P., 2009. Natural-historical diagrams: the 'New Global' movement and the biological invariant. *Cosmos and History: The Journal of Natural and Social Philosophy*, 5, pp. 92–104.
Woodard, B., 2013. *On an ungrounded earth: towards a new geophilosophy*. Brooklyn, NY: Punctum Books.
Wolfe, C., 2013. *Before the law: humans and other animals in a biopolitical frame*. Chicago: University of Chicago Press.

Resubjectification

6 Under heartbeat city's golden sun

Māori and the margins of performing the ultimate urban

Tawhanga Mary-Legs Nopera

When the vestiges of vacant anthills empty mouths of mother tongues, and the porous paupers of potbellied princesses parade facades of ferocity on TV screens, how do our broken hearts continue to beat? Bash and crash our lives onto concrete, metal tensions tire as our home-hearth fires dwindle like deadened sunsets. And yet, a golden-skinned hue of hope hearkens. If our genealogies are our bodies, they bare themselves to witness the warmth of bitumen and tar, far removed from the wharetangata, or womb, of our browned and ancient wisdoms. In trekking the mass urbanization of Māori people from traditional village nations, into warring global cities of cultural cataclysm, this research identifies the visage of neo-post-colony brethren. It seeks to stir the coals of the cityscape, the escape from Absalom's promised promenades to describe potential futures for people pushed to the margins of mediocrity.

Academic modes require degrees of distance and assume an impartial stance as a lens toward unhindered logic and cold objectivity. However, this chapter presents creative practice research as a means to empower multiple liquid realities within frozen concrete spaces. Māori scholar Carl Mika explains how discussions of Indigenous theory amidst those that birth from the West both veil and bring into being latent spaces of potential (Mika, 2013, p. 24). To awaken that which might appear dormant, this chapter locates the writer viscous within sites of seeming marginality; theory can both ensnare and unshackle the observer who is also subject. Creativity is a marginalized form of embodied academic inquiry, but one that can liberate covert forms of knowledge. This chapter makes an assertive stand for the logic embedded in emotion.

Within, I shift between subjective and objective positions, and emphasize words as abstract machines that close and open space (Deleuze and Guattari, 2004, p. 62). A shifting vantage gives voice to marginalized knowledge unique to the oppressed. Initially, ways that Māori are ostracized from mainstream Aotearoa New Zealand culture is described. Contemporary lines of flight are narrated as strategies toward tino rangatiratanga, or self-determination, but sovereignty is no simple strategy; liberatory outcomes can evade discovery until a critical mass of change tips the balance of power. In this chapter I offer creative knowledge as equipoise, a means to stabilize transcultural insecurity. Here, focus is given to friendship, partnership, and connective ways of being.

Finally, both the process of raranga and the concept of whakapapa are detailed as praxis. Raranga, the practice of traditional Māori weaving, is described as a research methodology grounded through its own knowledge foundation. Likewise, whakapapa, the Māori concept of relating, is discussed as Indigenous assemblage thinking. Māori scholar Leonie Pihama asserts, 'Whakapapa is regarded as an analytic tool that has been employed by our people as a means to understand our world and relationships' (2010, p. 5). Whakapapa relates Māori to every aspect of our reality and can reimage marginalized positions in globalized contexts. By assembling multiple forms of interpretation, this chapter explores the margins as providing agency for those who are bound within.

Navigating the noble savage

When Whitney Houston asked where do broken hearts go (1988), did she mourn for a broken heart, or imagine beyond the echo of its hindered beat? It's a feat to imagine beyond boundaries, but it is possible. In this chapter, I use the creative methodologies of Māori theory to emancipate captives from their subject-hoods; I theorize Māori futures as freed from capitals of colonialism. To begin, this chapter explores Māori ancestral strategies, which can give ground in the seeming fluidity of a globalized and virtual reality. Māori urbanization, rather than being described in its destitute form as a type of departure, is instead restoried; Maori subjectivity, as reterritorialized through urbanization, is the focus of this chapter. It seems like we've been smashed, broken, and beaten into an overtly oppressed outlook, but perhaps within our urban underworlds, the embodied patterns of our ancestors live. Through this chapter, I redescribe perceptions of Māori marginality and reorganize colonial rhetoric.

New horizons beckon, which explain Māori reality as connected and empowered through whakapapa; and in this way, Māori assemble relationships to places, objects, people, and deeds. Carl Mika explains 'whakapapa' as often under-translated to mean 'layers of genealogy', but this belies a 'concealment'; the layers of genealogy revealed through whakapapa more accurately explain the implicit relationship that Māori people have with the Earth and all things created by the universe's energy (2014, p. 54). Here, whakapapa is used to redescribe Māori urban marginality as a means rather than an end. I explore whakapapa as an assemblage of relationships, and locate myself within a community of Māori relations from Ohinemutu, a traditional Māori village that has become a marginal and yet crucial space within the city of Rotorua, in Aotearoa New Zealand. Within this chapter I occupy the positions of researcher and subject, insider and outsider, and explore how whakapapa helps make sense of a space that has slowly been marginalized. I describe how whakapapa orients bodies; it reassembles imposed geography and invigorates lost autonomy.

The purpose of this chapter is to reinterpret steel structures of social sedation and reframe them through perspectives of familial identity; I use raranga as a means to do this. Raranga is the Māori practice of over and under weaving, and as a framework for thinking, raranga can help bring hidden realities to the forefront.

'Maori frequently use the metaphor of raranga to explain a range of social and cultural processes that are integral to Te Ao Maori [the Māori world]' (Timu-Parata, 2010, p. 51), and although traditionally passed on through creative practices, it is the connective knowledge within raranga that affirms it as a research methodology. When I describe raranga within the context of this chapter, I imagine an empowered trajectory of urban realities for Māori; I contextualize this research through a partnership proposed between Te Arawa, a Māori tribe, and a local government body, the Rotorua Lakes Council. Raranga here is both a practice and theory, in that this chapter intends to weave ideas about marginalization with discussions on Indigenous and colonial forms of partnership. Through this chapter possibilities become pastiche; post-its of remembered pattern that wipe away the chalk-bordem of being at the bottom.

Theory chalk, Māori talk

If Western theory is a city, then it's a cold and snowy-white space, where those who are observed are often ostracized; its history asserts that impartiality allows objectivity, but sometimes that is just an excuse to exclude. At the lower levels of theory's towered territory, Māori have hunkered down, been trodden on, and tiptoed around. Underneath the cityscape, we have scraped for morsels, as experts from other cultures explain our existence from within their pinnacles of information; they seem to know everything about why we fail amongst their facades. The theory city has flayed us, by teaching us to forget the unique perspectives of Māori culture:

> The field of education is filled with theories – for instance, theories of child development, of learning, educational psychology – that have not been useful for us as Māori. In fact, they have been damaging to us, because underpinning those theories are deficit theories that position Māori as lacking, as inadequate and problematic.
>
> (Professor Graham Hingangaroa Smith quoted in Hoskins and Jones, 2012, p. 11)

We've been begging and borrowing to make ends meet, sweeping up what little is left of ourselves to make things right. If theory is a city, then its objective nature is no substitute for the warmth of inclusion; in Aotearoa New Zealand, I search for the clusters of community that signal my ancestors' vision of home.

Māori arrived in Aotearoa New Zealand with intent, with a purpose to relate to new territories, however, 'in the New Zealand context, the nation has rarely been imagined beyond the dominant colonial model, which focuses on the narrative of becoming New Zealanders. Māori may not have had the same understanding of "nation" as Europeans, but Māori still imagined a nation in many ways' (Mahuika, 2009, pp. 135–136). When the cities of the West began to reach our utopia, in the middle of the 18th century, slowly but surely all Māori hopes of belonging began to diminish, as centres of colonization replaced our potentials with ostracized peripheries. The hegemony of race inferiority became the foundation upon which settler identities removed Māori from the centre of Aotearoa New Zealand identity, and the ability for Māori to determine futures faded amongst the arrogance of scientific classification. The past, present, and future of Māori deficit has been meticulously crafted and mediatized through mainstream means; its intent is to deny us a continuation of presence in our own place. In Aotearoa New Zealand, it remains normal to publically reject Māori as belonging, as the following suggests:

> what on earth is racist about wanting New Zealand for all New Zealanders? Those New Zealanders that choose to label themselves something else (in the 21st century really? indigenous peoples inhabiting NZ – there are none left) over and above their nationality (which is an accident at birth) need to get a life and start to contribute to benefiting our overall society.
>
> ('Peter' – Online Comments, Rotorua Daily Post, 23 March 2015)

As a solitary sample, this comment reads as absurd, but on a daily basis, Māori are restricted through spite and relegated to refusals as a persistent colonial 'mine not yours' attitude alienates Indigenous people from our identities, land, and resources.

Since colonization, Māori have asserted both active and passive forms of resistance because rather than accept colonization as the norm, where Māori adopt strategies of resilience in response, '[r]esistance offers a much stronger set of political tools and is aimed more at dealing with the causal factors' (Penehira et al., 2014, p. 103). The refusal to remain silent has rendered Māori as radicals, renegades, and terrorists, where 'the politics of naming "terror" can be conceived as a biopolitical project aimed at governing Indigenous sovereignty' (Devadas, 2013, p. 5); media language has positioned Māori as the outspoken 'other', with the intent to tame, a means to make life better for 'ordinary' New Zealanders. But why aren't Māori ordinary New Zealanders too? Is it because we continue to work for a vision of country that predates a disenfranchising national identity? A democratic nation for Māori differs from the constructions of our colonial counterparts; for us, democracy is where communities direct their own realities, rather than be governed by a central bureaucratic structure, construed for the benefit of those with power. Misinformation about the ways Māori define identity are popularly promulgated by press-professors, who denounce kinship community models as exclusive and undemocratic because '[t]ribalism is based on principles of inequality. Democracy is based on equality. Kin status is what matters in the tribe; citizenship is the democratic status. Tribalism is exclusive. To belong you must have ancestors who were themselves born into the system' (Rata, 29 Jan 2013). But this ill-informed suggestion denies a paramount understanding within Māori culture: that all things in the universe are intimately connected and related; Māori social protocols encourage people to find ways to relate, ways to have voice, and ultimately, provide a means towards forms of self-determination.

The colonial city of smarts has successfully served to sever relationships, where healthy connections hide in urbane compartments individualized from community; in concrete cells of civility, commonly held goals are shaped through TV missives. Mass mediatized knowledge tricks and tempts totalitarian types of hype; '[t]he modern media assumed the role of the church, in a more secular age, of interpreting and making sense of the world to the mass public. Like their priestly predecessors, professional communicators amplified systems of representation that legitimated the social system' (Curran, 2002, p. 77). The theory city lays wasted terrain for communities of Māori because our ways of being contrast too sharply with the individualism central to consumer ideologies of control. Heidegger posits technology as an apparatus of production for future reference, stating that '[t]he revealing that rules in modern technology is a challenging . . . which puts to nature the unreasonable demand that it supply energy that can be extracted and stored as such' (Heidegger, 1977, p. 14). Here in Aotearoa New Zealand, Māori have been media techno-revised into chains of supply for someone else's net worth.

But what kinds of media have Māori brought from our pasts to remediate a reality so heavily encoded with an emphasis on another's culture? I posit that amongst the madness, Māori forms have been quietly emancipating; as abstractions of real-life they have bypassed colonialism and communicated connectedness. The koru (Figure 6.1), a simple unfurling spiral used as a basic element of Māori design,

Figure 6.1 Koru.

Source: Nopera (2015).

conveys layers of knowing and contemplation where, in a formal sense, the koru is proactive and growing; reflexive and looking back on itself; both inwardly and outwardly focused; as well as centralized and peripheral through constant outward movement. Yet, as an abstraction of a fern frond, the koru represents a deeper body of knowledge, where creation emerges from Te Kore – the void – and Te Po – the night – to birth Te Ao Mārama, or the world of knowing. The koru is a reminder that growth can come from nothingness; it invites potentiality. Moana Nepia explains Te Kore as a response to reterritorialization, which '. . . articulates how some of the effects of colonization, namely partitioning and surveying, can obliterate knowledge fundamental to Māori understanding of the world, leading to frustration, trauma and anger' (Nepia, 2012, p. 71). However, the relationship between the koru and Te Kore

> may also articulate a continuum, for the layering of views. As eternity, Te Kore articulates space in which we may speak and move, or be denied opportunities to express ourselves. Moments of calamity, or uncertainty when all seems disconnected, and unsuitable must be overcome if a creative journey is to fulfil its purpose.
>
> (Nepia, 2012, p. 70)

Māori media have significant power to transform and liberate spaces that seem marginal, unwanted, and without value because they are embedded with 'the good stuff'.

For Māori, these externalized aspects of culture have allowed adaptations, and this is something asserted by Penehira, Green, Smith, and Aspin when they explain, 'Māori living in cities and in other parts of the world continue to draw upon Māori values and traditions, albeit refashioning and employing them so as to foster and maintain their well-being outside of their homelands' (2014, p. 98). Since colonization from the West, appropriation of our forms, objects, and knowledge has gained Māori a type of global coverage. On the one hand, our things have been thieved away, or cheated from us through mistrustful deeds; and yet those same tourist-type transactions have provided us with our own knowledge in alien lands. The structures of theory embedded in city spaces have enacted forms genocidal violence because although the harvest of history provides a basis for Indigenous identities, they will always be ghostly; it takes our losses to manifest museum mementos. The city is a consumptive beast because it eats relationships, social engagements become transactions of power, competitions to see who can take the most. Maybe somehow, being Indigenous can unsettle an urbanized modus, simply through being contradictory?

Hundreds and ones, joining the dots

The urban realities of marginalization imposed upon Māori people result in continued land loss, failing educational outcomes, addiction, a forgetting of language and cultural patterns, incarceration, and premature death by suicide or chronic illness; intersecting tyrannies are casting Māori as immobile captives, caught once, twice, and so on forever. At 39 years of age, the narrative of my own life has already been adversely affected by the layers of marginality mentioned, which have been helping me towards a self-imposed or early end. Māori educational theorist Jenny Lee (2009, p. 2) writes, 'Storytelling has always been one of the key ways knowledge was sustained and protected within Indigenous communities', and from this point on, I deliberately share story as a way to prevent my words, and ultimately myself, from becoming trapped in the graveyard of theory's city. Lee theorizes about storytelling, or pūrakau, from a Māori worldview:

> Pūrākau provide a conceptual framework of representation that is relevant to research. The innovative methodological work of international Indigenous scholars as well as local Māori writers and academics provides inspiration to look beyond conventional research methods and academic styles of documentation and return to our own narratives, to experiment with literary techniques to research, and disseminate knowledge in ways that are culturally relevant and accessible.
>
> (Lee, 2009, p. 5)

And so, a story about friendship begins.

Friendship might not seem the most appropriate topic for discussion here, but really it is the kind of relationship that might best give sense to ideas discussed in this chapter. Primarily, I shape this story around my friendship with Unangax (Aleut) Anna Laura Hoover, an artist, fisherwoman, community activist, academic, and all-round incredible person. I met Anna in 2007, at Piko, an Indigenous artist gathering held in the highlands of Waimea, on the Big Island of Hawai'i. I don't know how I ended up there because I had only just finished art school six months beforehand; finally after more than a decade in undergraduate study I had completed a course of academic study, to end up making art alongside the most prominent Indigenous artists in the world. Anna is one of those enigmatic people other people enjoy being around, and over the eight days of the gathering she was one of the people I enjoyed spending time with. At that time it was a brief introduction to friendship because immediately after Hawai'i, I travelled to Vancouver, BC. I mistrust that city, not for its beauty, but for the aching sense of isolation I felt living there. In the year that I lived there, I had never experienced a city so surrounded and filled with beauty, that at the same time felt empty and without aroha, or love.

I contracted HIV in Melbourne when I was 22 years old, after my father's premature heart-attack death fuelled a sense of loss that made me wish for death; ten years later, Vancouver was the city where I thought my ancestors would finally grant that destitute desire. I was shocked in the isolation of a concrete urban space so colourfully decorated with Indigenous forms, where Indigenous people and Indigenous spirit felt as if they had been all but silenced. I really missed home that year. I don't know who, but someone back in Aotearoa New Zealand told me about an exhibition I should go to, showcasing a collaboration between Māori pounamu

(Jade) artist and a Tlingit glass artist. If anything, it was an opportunity for me to see people from home and perhaps network myself into the First Nation's art scene that I knew existed but just could not find. And there, between courses of expensive food and vodkas paid for by Spirit Wrestler Gallery, I got to know Anna Hoover much better; if I hadn't seen her that night my life most probably would have ended in Vancouver because the helplessness of ongoing poverty and the fear of homelessness had begun to spin webs of suicidal self-talk.

Anna and I, two seeming strangers at the time, organized to catch up across the Canadian border in Seattle, as Anna's family live between Southern Puget Sound and Egegik in Bristol Bay, where they fish commercially for wild sockeye salmon. We got on like a couple of crazy art inspired peeps, and her friendship at the time really helped guide me back to spiritual health and then towards further academic study. At the beginning of 2008, still based in Vancouver, I began a master's degree in fine arts through an Aotearoa New Zealand institution, while Anna went a step further to concurrently study towards two master's degrees, in Native American art history and also Indigenous documentary filmmaking. After a few months I moved back to Aotearoa New Zealand to fully immerse myself in art study, but Anna and I kept our connection strong, writing to each other daily via Gchat and Facebook, both of us trying hard to navigate the cold and exclusive streets of a theory-space. Art helped make sense of all the theory, although the theory seemed so far away from my lived reality; everything I read reiterated the cement of oppression and it was demoralizing to know how fixed in words my world was (and still continues to be). However, in those Gchats and Facebook posts from Anna, I could clearly read the over-arching nature of inequality for Indigenous peoples; from the other side of the globe, Anna's experiences of theory and Indigeneity were mirroring my own.

As part of Anna's final work for her master's in Native American art history, she organized a touring exhibition of t-shirt prints, created by Indigenous artists from around the world. Titled *fashion STATEMENT: Native Artists Against Pebble Mine*, the show's intent was to draw attention to a gigantic mine proposed for the headwaters of Bristol Bay. As one of the contributing artists, I created an image to voice my concern over the potential destruction of an entire ecosystem; the proposed mine could drastically end the subsistence life-culture perpetuated for thousands of years in the region. At the same time, I also wanted to communicate similar kinds of propositions made in Aotearoa New Zealand, and assert how relationships with Indigenous cousins around the world enable greater resistance, strength, and voice. My contribution, titled ANGER DANGER (Figure 6.2) depicts a single line of communication between Anna, on the left, and myself on the right; the line between us is interrupted so that it becomes jagged and uneven, appearing as ocean waves between Aotearoa New Zealand and Alaska. At the crest and in the trough between each wave there lies a koru; as each part of the line is broken and ends, something new transpires, and so on until a meaningful and lasting connection is made. This is the intent of whakapapa; as a process of relating all things in the Universe, distant assemblages are layered into new dreamscapes, and thus, the Universe grows.

Figure 6.2 ANGER DANGER.
Source: Nopera (2010).

In simple terms whakapapa translates as *whaka* – 'to be' or 'being like'; it turns a noun into a verb, and *papa* translates here to mean the Earth; so whakapapa is the process of making people be the Earth. Whakapapa is an embodied form of genealogy that connects people to each other, to animals and plants, to all of the energies of the planet, and the Earth as part of an expansive cosmos; but it isn't a singular system or framework that Māori apply to the world because whakapapa is the intent, the theory, the practice, and the outcome. A way to consider whakapapa could be to imagine a connection; whakapapa is the desire to interpret and acknowledge that connection; whakapapa is the connection itself; whakapapa is the pūrakau, or the story of that connection; and ultimately, whakapapa becomes the desire to interpret and acknowledge any new set of connections that are to grow from the connections already made. In the theory city, rather than multiply and expand the Universe, assemblages from the West have broken peoples' perceptive abilities into fractional chunks, removing their ability to simply relate. Even when Māori were almost marginalized to extinction, the profundity of whakapapa kept us from forgetting ourselves because whakapapa continuously imaged us as assemblages that include even our oppressions.

Whakapapa inspires life; it allows Māori people to keep memories alive, to reorder them and test out their relevance over aeons through restorying histories; through whakapapa, we are able to link and connect to things that seem distant and abstract, as well as those that seem close and more real. Through the theory of whakapapa, people are always able to locate themselves at the core of their accumulated experiences; thus, whakapapa resists marginalization by centring identity.

Just weave me alone?

It may seem that whakapapa is just what people do when they get to know each other, and in some ways this is true; however, over time Māori have developed our own unique protocols and traditions to enable smooth social transactions; amongst a nation of tribes, each with our own political intents and tribal aspirations, it has been crucial to develop patterns of engagement to protect the diversity of knowledge and resources available. A person could spend a lifetime getting to know the ways of any one tribe and still only scratch the surface of its whakapapa; before colonization, tribal experts were able to recite the names of every ancestor since creation, when the Earth Mother and Sky Father were separated by their children. Like that primordial space of creation, the margins are cramped to refugee capacity, and I ask here, in the theory city, how can people use whakapapa as a means to reformulate and rectify imposed oppressions? I suggest that the process of raranga, or weaving, can help. To me, friendship is like weaving because it ends solitude through the creation of something new; it begins with an intent that changes to an acknowledgement, it shifts from gathering materials to preparation, from grounding to creation, and it is formalized through resolutions, more acknowledgements and further intent. Friendship, like the koru, is circular and growing, but it is not always easy to think in circles when theory cities have conditioned us into segmented hierarchies; here, raranga becomes helpful because it is circular and hierarchical in the same instance; at once it is cyclical and linear, and in this way, raranga provides a multidimensional means towards the affirmation and creation of assemblages.

I was raised in a small village named Ohinemutu, in Rotorua, where the majority of my ancestors moved after wars with another tribe nearly decimated us; although, as with all seeming ends, the warring formulated new whakapapa where we became related to our enemies. After that time, when the en messe settlement of Aotearoa New Zealand from Europe began from the middle of the 1800s, even newer forms of whakapapa began to develop, and with the intent of partnership the settlement of Rotorua very quickly became a centre of shared aspirations between Māori and Europeans. Dreams can be hazy, and in the madness of colonization, craziness certainly ensued; Māori had ways of doing things that conflicted with theories promulgated by the settler culture and the vision of friendship became as vague as a dream. As a child and young adult I heard the stories about the vast tracts of land that were gifted or sold by my ancestors, or confiscated to create space for Europeans; I have held on to a lot of anger knowing that good faith was quickly forgotten, by those who arrived and benefitted from acts of generosity and theft. By the same token, there grew a degree of resentment by settlers who often felt excluded from the relationships with land that Māori had already formed; the imported political framework that forced citied ways of living upon Māori refused to allow positive connections and ways to relate to this new space for settlers too. Nearly 180 years later, the emptiness of that experience is being resolved in quite empowering ways; and I argue here that these new forms of democratic exchange are allowing us to raranga, or weave, new whakapapa and ways of being.

Iwi are political groupings of tribes with shared genealogies, and in Rotorua, Te Arawa is the confederation of tribes who have tied our identities to the district's resources over a long period of time. In 2014, the Rotorua Lakes Council sought a way to better partner with the Iwi of the Rotorua region, after decades of broken-down consultations led to intense legal actions and significant financial losses for the Council and ratepayers. The intent is obvious: a formal partnership between peoples who have inhabited a space for more than a thousand years and the new community formations that have grown since colonization. However, for this to happen acknowledgements must be made that allow people ways to communicate, both the positive and negative things that have either encouraged friendship or denied its relevance, need to be exposed. From 2014, Te Arawa people began to describe these aspects in an attempt to configure a pathway; a working party moved through the various tribal areas within the district to collect stories and histories, to create a potential model for Iwi partnership with local government. The year-long process exposed a huge degree of hurt and pain; not just those of Māori experiences, but also the sorrows, concerns, and fears of European descendants living in Rotorua which emerged in response to partnership discussions. It was a difficult time because my own feelings about misdeeds enacted upon my ancestors made it hard to understand the anxiety about a partnership with my Iwi; all I could hear around me and read in the paper were arguments that excluded me from belonging to the emerging nation, and I felt a lot of racism during the initial consultation process.

However, it is important to expose feelings to decipher their origins, and in the very publically mediatized acknowledgement of shared histories of neglect, a form of harvest took place. We call it public consultation, but it is a harvest nonetheless, where thoughts, perspectives, ideas, and stories are gathered. If people want to make friends with each other, they have to honestly gather about them the social and cultural offerings being made; to make friendship happen there has to be truth, even when it brings discomfort and distress, otherwise friendships will remain as fictions and frictions. The consultation process opened old wounds through which people bled words, to each other, to the district council, and to Te Arawa; when intents have become blurred reality festers in the disjuncture. Preparation must happen when an important objective is to be realized; realness enables sense:

> In such conflicts, it is not the planning *per se* which is the problem. However, the fact that the system becomes an outlet for the deep-seated fears, aversions,

> or anxieties of some residents means that this does become a problem for the planning system, in terms of responding in constructive ways.
>
> (Sandercock, 2000, p. 16)

By harvesting ideas and words and laying them out so that they are visible to all, people can collectively sort constructed thoughts and in this way interpret potentials; preparation makes room for work to begin.

As an artist, I like to believe that I am clever enough to make on a whim and often that's exactly what it seems like I am doing; the truth is, however, no work can happen without an intent, an acknowledgement, a harvest, and some flavour or preparation. The embodiment of a process can do this; it can cause me to forget the lifelong work that it has taken to grow an idea, or to create a relationship; when people decide to make things happen they ground ideas through their actions – grounding, to be like the ground; whakapapa, to be like the Earth. Whakapapa happens when a person activates relationships through bodies, and with regard to the Te Arawa Partnership Proposal, whakapapa implements the design and integrity of all the preceding work. The model devised by Te Arawa, with support from the Rotorua Lakes Council, became grounded as it manifested through many conversations; emotions and stories carried over aeons of time by Te Arawa and settlers alike helped grow a model for partnership. On 26 May 2015, the Rotorua Lakes Council voted 8:5 to implement the Te Arawa Partnership Proposal. As I listened that day, to Councillors debate on behalf of the various assemblages they represented, I felt emotional and proud to see people raranga thousands of years of knowledge from all over the world, into a decisive action plan for future generations.

Many people in Rotorua feel marginalized, not just Māori, because we live in a beautiful city abundant with resources, and yet something about its theories has halted us in our ability to fully enjoy our lives. The affirmation of whakapapa within a raranga process is only partway along the journey because a whole new form is yet to be created and resolved, acknowledged, and interpreted for whatever vision comes next. What form or ideas might come from this partnership model between Māori people, and the theories that have shaped our settler friends and kin from the West, is difficult to know, but I sense something beautiful. I remember being in a Paris nightclub once, where a young woman came up to me on the dance floor, grabbed my earring and said, 'This is not beautiful'. At the time I didn't care because I really felt I was rocking a great look; she and I didn't have the time to talk about our different ways of beholding beauty and I still sometimes wonder whether it was me or the earring she was describing. Raranga isn't easy; it is slow, it takes focus, and it requires space and time to share as disparate strands formulate into something not yet in existence; like the koru it asserts that where it was empty, it can be both full and beautiful.

The golden sun

Rangimārie Mahuika describes Māori theory as allowing 'an assertion of our cultural beliefs and practices, our ways of knowing and being and our right to both live and maintain them' (2008, p. 4); this chapter has emphasized story as means to theorize in Māori ways. Rather than endorse the isolation of the theory city, I have attempted to write about Māori ideas so that they can be interpreted from the city's margins, which is pretty much everywhere; I don't want Western theory to define my life anymore. The lens of Western theory tends to subjectify me in my life, to the point where I am a puppet without my own dreams, desires, and destinations; the spaces of those theories are like a desert city, where although there is sun, there is also an ongoing thirst for reprieve. In the dry and dusty theory city I am no longer myself; instead, there I am a corpse of words written by someone who doesn't really know me but knows they need me to play along. Western theories are interesting in this regard because they seem like they might be for the marginalized, as a way to advocate for betterment, and yet so often in truth, they are for the betterment of theorists. In writing this chapter I have wanted exactly the same thing: the betterment of myself as a theorist – the difference is that as well as the theorist here, I am also the subject; subjects have their own theories, and if we can find ways to hear and see these, theories might be able to turn into emancipatory practices. Diné/Taskigi artist and scholar Hulleah Tsinhnahjinnie explains that '[t]he colonized perspective is the perspective that is from the boat . . . [w]hat I try to show in my images is a perspective from the beach of how we see that boat: an indigenous perspective with indigenous references and touchstones, not Western-based analysis' (Tsinhnahjinnie quoted in Schneider, 2013), and in the same sense, I attempt to use my own grounded space to explain what I see and experience.

Whakapapa has been described within this chapter through use of narrative, where the ability to relate is analogized in a story about friendship. Connections are important because through exchange, the value of shared experiences can help bind distant territories; whakapapa can help to reassemble imposed geographies so they make sense to those who occupy them. Whakapapa is assertive; it resists colonial marginalization through clear talk, meaning making where people communicate about ways in which they might be able to relate. Although we are all unique in this world, connective strategies can help us create bridges between our assemblages of reality, and those of people who might seem quite different; whakapapa can help communities see beyond the names that define them, to imagine new words and contexts, where margins become abstracted to represent unseen and dreamy terrains. Deleuze and Guattari's theories about assemblages, although empowering, are locked in words that perpetuate long-held power relationships of oppression; where the theory city continues to dominate through its form of detached storytelling, academia stories a status quo. By considering its notions from a marginal perspective, this chapter demonstrates ways that abstract knowledge can become real; through Indigenous lenses and ways of being where theory is embodied, Deleuze and Guattari's ideas can be translated in new ways, as hope for our collective futures.

Finally, this chapter describes raranga, an ancient process practiced by all cultures of the world, although here, it is specifically detailed in Māori ways; I only know how to weave as a Māori person where within my culture, raranga reinforces the importance of people within families, within communities. In this chapter I have described a process of partnership between my tribe, Te Arawa, and the local government body which through colonization now directs our lives; I don't like having other people govern my life, but that is a reality I must deal with in today's world; if I can find a means to govern myself at the same time then I can be happy. The Te Arawa partnership with the Rotorua Lakes Council is history making because never before has a colonial body actively sought to give voice to Indigenous governance structures. My ancestors resisted colonial oppression for close to 180 years, and now, the positive contributions Māori might make toward a partner-identity can move beyond tokenism towards truer forms of democracy and self-agency. I acknowledge here that this is only a beginning, but as I detailed in descriptions of the koru, nothings allow beginnings to grow into new understandings and good ways forward. For me, the ability to whakapapa through raranga, or to ground myself through weaving, helps me move beyond the bounds of a lived urban marginality.

References

Curran, J., 2002. *Media and power*. New York; London: Routledge.

Deleuze, G. and Guattari, F., 2004. *A thousand plateaus*. London: Continuum.

Devadas, V., 2013. Governing indigenous sovereignty: biopolitics and the "Terror Raids" in New Zealand. In: B. Hokowhitu and V. Devadas, eds., *The fourth eye: Maori media in Aotearoa New Zealand*. Minneapolis: University of Minnesota Press, pp. 3–24.

Heidegger, M., 1977. *The question concerning technology: and other essays*. New York: Garland. pp. 3–33.

Hoskins, T.K. and Jones, A., 2012. Kaupapa Maori: the dangers of domestication. *New Zealand Journal of Educational Studies*, 47(2), p.10.

Houston, W., 1988. *Where do broken hearts go* [lyric]. Nashville: Arista.

Lee, J., 2009. Decolonising Māori narratives: Pūrākau as a method. *MAI Review*, 2. Available at: <http://www.review.mai.ac.nz/index.php/MR/article/view/242/268> [Accessed 19 Jul. 2013].

Mahuika, N., 2009. Revitalizing Te Ika-a-Maui MĀORI migration and the nation. *New Zealand Journal of History*, 43(2), pp. 133–149.

Mahuikia, R., 2008. Kaupapa Māori theory is critical and anti-colonial. *MAI Review*, 3(1). Available at: <http://www.review.mai.ac.nz/index.php/MR/article/view/153/180>. [Accessed 16 Aug. 2014]

Mika, C., 2014. The enowning of thought and whakapapa: Heidegger's fourfold. *Review of Contemporary Philosophy*, vol.13, pp. 48–60.

Mika, C., 2013. Western "sentences that push" as an indigenous method for thinking. In: A. Engels-Schwartzpaul and M. Peters, eds., *Of other thoughts. Non-traditional approaches to the Doctorate*. Rotterdam: Sense Publishers, pp. 23–26.

Nepia, P.M., 2012. Te Kore: exploring the concept of the void (2). Ph.D. Auckland University of Technology.

Nopera, T.M-L., 2010. *ANGER DANGER*. [t-shirt print] (fashion Statement: Native Artists Against Pebble Mine. Anna Hoover Collection).

Nopera, T.M-L., 2015. *Koru*.[digital image](collection of the artist).

Penehira, M., Green, A., Smith, L.T., and Aspin, C., 2014. Māori and indigenous views on R & R: resistance and resilience. *MAI Journal: A New Zealand Journal of Indigenous Scholarship*, 3(2), pp. 96–110.

Peter, 2015. *Online newspaper comment* [Rotorua Daily Post]. 24 March. Available at: <http://www.nzherald.co.nz/rotorua-daily-post/news/article.cfm?c_id=1503438&objectid=11421789> [Accessed 24 Mar. 2015].

Pihama, L., 2010. Kaupapa Māori theory: transforming theory in Aotearoa. *He pukenga korero,* 9(2), pp. 5–14.

Rata, E., 2013. Elizabeth Rata: tribalism, democracy incompatible. *New Zealand Herald* [online] 29 January. Available at: <http://www.nzherald.co.nz/nz/news/article.cfm?c_id=1&objectid=10861949> [Accessed 15 Feb. 2013].

Sandercock, L., 2000. When strangers become neighbours: managing cities of difference. *Planning Theory and Practice*, 1(1), pp. 13–30.

Schneider, W., 2013. Photo essay: Hulleah Tsinhnahjinnie. *Cowboys and Indians: The premier Magazine of the West,* March. Available at: <http://www.cowboysindians.com/Cowboys-Indians/March-2013/Photo-Essay-Hulleah-Tsinhnahjinnie/> [Accessed 22 May 2015].

Timu-Parata, C., 2010. Making a parallel between raranga and undertaking a literature review. *Whitireia Nursing Journal*, 1(17), pp. 50–51.

7 'The ghetto will always be my living room'

Hustling and belonging in the Nairobi slums

Tatiana Thieme

Hustling on the margins

This is the story about the journey of a young woman who grew up in one of the oldest and largest informal settlements of Nairobi. Eliza's story is inextricably tied to the broader narratives of *hustling* that feature centrally in everyday 'creolised argot' (Comaroff and Comaroff, 2005, p. 28) amongst Nairobi youth making a living outside the formal economy. 'Hustling' in the Kenyan context has shaped youth identities and their norms of adaptation to everyday adversity and uncertain futures. Eliza was one of the key protagonists of my PhD research, in many ways epitomizing and defying the logics of the 'hustle economy' (Thieme, 2013). Her journey included the impasses she faced as she aspired to *do well* – to get out just enough to be better off – but to stay anchored within the collective subculture of urban youth living in informal settlements or 'slums'.

This chapter is situated within efforts to theorize youth agency and struggle 'in the hood' by drawing attention to the economic, cultural, material, and political repertoires of largely unheard voices on the urban margins (Honwana and De Boeck, 2005; Jeffrey and Dyson, 2008; Richardson and Skott-Myhre, 2012). Increasingly, youth contest social and environmental injustices by connecting their marginalized status as formally *jobless* youth to unoccupied terrains of potential income generation to assert their place within their neighbourhoods and within Nairobi more broadly. Take the small-scale but widespread youth-led operations of waste work that have been taking place across slums of the city, born out of the city's gradual physical infrastructural decay since the 1980s combined with a widespread accepted informality in nearly all spheres of urban service provision (Myers, 2005; Huchzermeyer, 2011).

Eliza grew up in one of the most congested low-income estates of Nairobi known as *Huruma* near one of the largest and oldest informal settlements (see Figure 7.1), Mathare Valley (Etherton, 1976; Chiuri, 1978). She was amongst the first generation of young Kenyans in the post-colonial capital who were born and raised in what most young people call the 'ghetto', and who were left to their own devices to make a living. The challenges and everyday struggle in the slums should by no means be romanticized, but it is worth emphasizing that *the hood* has become generative in its own right, a context where traditional cultural norms and roles can be challenged,

Figure 7.1 Mathare Valley rooftops.

Source: Tatiana Thieme, 2009.

and where the regulatory and legalistic parameters of urban provisioning could be pirated and creatively circumvented (Sundaram, 2010; Mabala, 2011). The informal waste economy has placed formally invisible and ill-served pockets of Nairobi on the radars of multiple actors across government, business, and NGO sectors (Thieme, 2010). The celebration of the informal sector in Kenya has raised polarized debates concerning the justification of state support retrenchment versus the normalization and expectation that all the poor must be entrepreneurial subjects in order to survive, let alone thrive, in the harsh urban environment (Hart, 1973; King, 1996; Potts, 2007). But beyond the debate on state versus market mechanisms lies the subjectivities of these 'entrepreneurs' themselves that have been, to date, rather under-documented, particularly in the case of youth, and especially female youth.

Numerous urban sociologists have theorized urban marginality by focusing on the 'ghettos' of the US and Europe, highlighting the post-industrial economics, racial demographic segregation, and politics of welfare state retrenchment that have continued to further exclude urban 'outcasts' from mainstream urban society (Wilson, 1987; Comaroff and Comaroff, 2005; Wacquant, 2008). Similarly, youth studies have become a growing area of academic and policy interest, especially on issues of unemployment and the dangers of mass youth disaffection, portrayed either as the 'ticking time bomb' or the source of hope and innovation (Diouf, 2003; De Boeck and Honwana, 2005; Jeffrey, 2012). While these are crucial contributions to the study of urban marginality, less attention has been paid to the internal cultural and material logics that inform everyday rationalities and experiments in precarious urban life (Gidwani, 2001; Simone, 2010). These include existing coping strategies of youth whose life, work, identities, and place within cities is often stigmatized and defined by negation, categorized as unemployed, idle, or a perceived nuisance for 'hanging about', loitering, or making trouble (Borden, 2001; Weiss, 2009; Jones, 2012; Thieme, 2013). Young women's stories and their own paradoxical subjectivities merit particular attention (Homwana and De Boeck, 2005). They occupy the liminal and politicized status of 'youth', whilst also epitomizing the everyday negotiations with gendered expectations associated with domestic responsibilities, traditional land inheritance norms, and stigma associated with teenage pregnancy and single motherhood.

The study of *hustling* builds on conceptualizations of the nexus between urban poverty and youth, reflecting everyday struggle in the face of extreme urban social environments and prolonged precarious conditions. Ethnographic accounts depict youth in various geographies navigating uncertain futures by engaging in 'modernity bluffs', finding creative channels to express disaffection, and shaping a kind of 'zig zag capitalism' based on the ingenuity of forced adaptation to rapidly changing cityscapes (Vigh, 2006; Waage, 2006; Newell, 2012; Jeffrey and Dyson, 2013; Jauregui, 2014; Fredericks, 2014). What has not been explored in depth are the internal frictions between those who stay 'stuck' in the poverty trap and those who manage, albeit in incremental ways, to do well, but whose subjectivity and sense of place is anchored in *the hood*, to the extent that negotiating everyday social life, work, and place renders hustling a performative practice. Hustling, for Eliza, involved moving in and out of the hood at various stages of her formative years, and managing the social ties that made these streets her *base* but also at times a potential trap. Eliza's story sits within broader post-colonial critiques of urban studies, by rethinking and theorizing subjectivities of marginality *from the South* (Myers, 2011; Comaroff and Comaroff, 2012; Robinson and Parnell, 2012), and more specifically from the *hood* where adversity and frustration are co-constitutive with alternative coping strategies and learning that shape particular experiences and attitudes (McFarlane, 2011; Richardon and Skott-Myhre, 2012). Understanding how slum dwellers navigate uncertain futures requires a critical engagement with the very expressive articulations our interlocutors deploy in everyday parlance, harnessing the everyday argot as potential analytical repertoires of critical urban studies. It is in this spirit that this chapter tells a story of hustling on the margins, seeking to emphasize the point of view of someone on the inner edge of the margins.

Ethnographic encounters

This chapter is an exercise in ethnographic portraiture (Jeffrey and Dyson, 2008), seeking to privilege the multitude of insights, the simultaneous impasses and possibilities and paradoxical subjectivities of a singular informant. Eliza's story is set within the broader context of an ethnographic study on Kenyan youth's diverse everyday strategies and experiences in navigating uncertain urban terrains, and how some youth have managed to appropriate their marginalized and liminal status to render 'the ghetto' a place of possibility. Elsewhere I have located the epistemologies of 'hustling' within everyday banter, accounts of life on the streets, articulations of struggle, work, and aspirations. These accounts emphasize the interpersonal dynamics and social ties that are integral to the hustle in the hood, which paradoxically are consistently anchored in the collective unit of the 'group', but sometimes erode solidarity and possibilities to *do well* (Thieme, 2013). While the *hustle economy* displays modes of improvisation and adaptation that engage with capitalist relations and entrepreneurial urbanism, it does not necessarily equate individualistic, self-interested pursuits of continuous and limitless economic gain, or desires for differentiation (Thieme, 2015). The hustlers' skill is instead to strategically modulate between survivalism, livelihood strategy through diversifying sources of opportunity to mitigate risk, and contestations of authority. The tenuous balance between feeding themselves and renegotiating their place within the city is at the core of urban hustlers' subjectivity.

Eliza's story told below is based on our countless private conversations recorded during my homestays with her and our numerous walk-abouts in her neighbourhood between 2009 and 2012, a series of detailed life history interviews conducted in 2010, and the informal group bantering between Eliza and her male friends with whom she had grown up, played football, worked as a garbage collector, mentored, and eventually bossed around as a sanitation entrepreneur and team leader. The story is largely told through my narrative voice but seeks as much as possible to have the palimpsest of her words, expressions, and experiences come in the foreground, as quotes from various recorded conservations are interwoven into the text. This chapter seeks to reflect the meaningful though inevitably partial encounter of ethnographer and interlocutor. It offers an assemblage of academic reflections concerned with the core theme of this book on urban marginality, by emphasizing the interpersonal palpability of the field in all its fragility and serendipitous experimentation through a mosaic of exchanges, where Eliza let me into her life world, and allowed me to ask her to articulate things that she had never been asked to reflect on before.

Eliza was in her early twenties when we met, and she embodied the paradox of youth growing up in Nairobi's urban slums. Her story as a single mother who got pregnant before finishing secondary school was typical in these low-income neighbourhoods, where countless teenage girls ended up having their youth prematurely truncated by unintended pregnancy, and the social stigma that followed the experience of being a young single mother. These two factors, single motherhood and the inability to afford living on her own, exemplified the *protracted liminality* of youth living in urban poverty in Nairobi (Thieme, 2013). On the one hand youth were forced to 'grow up' quickly and assume adult responsibilities; on the other hand the lack of job opportunities and ability to afford housing typified youth's inability to attain certain key markers of adulthood such as a secure source of income and being able to afford 'your own place'. This was all the more significant for young women who were the *de facto* primary carers of their children, as well as often dutifully tending to their elders.

At the same time, Eliza had defied the stereotypical ripple effects of single motherhood and youth unemployment. She had *not* dropped out of school when she got pregnant, instead finishing her secondary school against the odds. And she challenged gendered norms of her local entourage by inserting herself in the male-dominated informal economy of garbage collection that had become one of the main sources of income generation for young men 'in the hood'. Eliza always felt more comfortable hanging out with the guys. She boxed, she played football, she was a garbage collector, and she was able to partake in the macho banter with her male peers but in a way that playfully ridiculed their ways whilst appreciating their highly sexist humour. For at the end of the day, Eliza would never be one of those women that her friends referred to, the mistress or 'girl on the side', or '*Mpango wa kando*' (spare part) as the expression goes, but nor was she the girl that the guys courted either. She was their friend, their sister, in some instances their boss.

But as a young single mother, she also embodied the plight of young women living in the slums, knowing what it was like to navigate uncertainty and adversity in a neighbourhood where basic necessities like finding a clean safe toilet were a security concern, and where countless teenage girls who became pregnant had slept with men in order to get some money for food, phone credit, or even sanitary pads. Eliza related to her female counterparts especially when she spent hours at the hair salon getting her braids done every few weeks, and retained within her *garcon manqué* ways a modicum of *coquetries* and female complicity that gave her a foothold amongst the women in her neighbourhood. It was this dual position as a single mother who understood what it was like to grow up and raise a child in the ghetto, and her role as a youth female activist and waste worker hustling on the margins of the informal economy that accorded her a unique street credibility amongst both her female peers and her male peers.

Life in Huruma's hood

Eliza was born in Huruma in 1987, where she spent her childhood and early adult years. She had four brothers and one sister, all born in Nairobi. When I met Eliza in 2009, she lived with her parents, one of her brothers, and her son in a two-room flat on the bottom floor of a four-story tenement walk-up in the middle of Huruma, one of the most densely populated low-income 'estates' in the world, adjacent to Mathare Valley (see Figure 7.2). One room consisted of a double bed for her parents, a small coffee table sandwiched between the bed and a two-seater sofa, which turned into a makeshift bed for Eliza's seven-year-old son, and where most of the family assembled to watch television on a 12-inch screen perched on a tired-looking wall unit that marked the spatial separation between the 'bedroom' space and the 'living room' space. The second bedroom had one single bed for Eliza's older brother, and a mattress on the floor against the iron sheet wall for

Eliza. Between these two rooms, a small separate utility space was used for cooking and storage. The toilet, bathing stall, and water tap were down the hall and shared across all the families living on that floor. Although this part of Huruma was electrified, blackouts were a daily occurrence, and little pieces of waxed candlesticks were seen in most corners of homes and the common washing area, serving as makeshift lights after sundown.

As was typical for most children born in Nairobi's low-income settlements in the 1980s, Eliza's parents were from one of the rural provinces in Kenya and had migrated to the city for the father's work. At the time, he was a government employee, and her mother sold second-hand clothes. As Eliza explained during one of many recorded life history interviews,

> Eventually my mother's business collapsed. My childhood was not bad because when I was being brought up, my dad was still working. But by the time I was thirteen, he stopped working. I don't know what happened to the job. That is where life became difficult.

Although Eliza's father had been out of work for years, her father's former status as a government employee was an important part of the family folklore and pride. The second-hand clothing market, in which Eliza's mother was a 'businesswoman', was a large but saturated share of the informal street vending economy in Nairobi, and it was common for individual entrepreneurs in the second-hand

Figure 7.2 Room with a view.

Source: Sasha Turrentine, 2010.

market to give up their business at one stage or another, due to the significant logistical and economic challenges associated with inventory space, vending stall rental cost, and keeping up with local competition. When both her parents were out of work, Eliza's oldest brother, who had just finished secondary school, 'took care of the family'. When she spoke about this period in her childhood, she evoked a sense of gratitude towards both her brother and the benevolence of her primary school, as though inferring that this was unusual fortune within the slums.

Eliza's brother had become the sole breadwinner of the family by the time he completed school. It was not uncommon for a young member of the household to take care of the whole family in these neighbourhoods. What was less common was young people finishing secondary school, partly because this part of Nairobi had very few secondary school options, and unlike certain benevolent primary schools, most secondary schools were fee paying and did not accommodate for financial aid. Eliza had spent her early teenage years with her brother's example, hustling in the informal economy, generating income wherever an opportunity arose, and as long as his parents were out of work and younger siblings were in school, it was a given that the eldest child's obligation was to provide for the rest of the family. Eliza was influenced by her brothers and older boys around her neighbourhood in other ways as well. By the age of 15, Eliza's social and recreational activities were often spent hanging about with an array of 'big brother' figures who taught her how to train, how to be tough, how to defend herself. As she put it,

> My best childhood memories were playing football and boxing. I was usually training mostly with boys. My brothers were all boxers. So I was just going with them to see them train, and then I started training with them too. The women didn't think it was strange, because they knew I grew up with only brothers, so they thought I was just more comfortable with the boys. But the boys they were cool with me because they saw that I was not like the other girls.

In 2003, Eliza joined a nearby secondary school. But within a few months, Eliza realized she was pregnant and dropped out of school. The tone of her voice and facial expression changed suddenly as she recalled,

> After that life became difficult. My parents weren't very happy with me. My older sister was not really around because she is married. You know, the situation was such that I was the last born, and now pregnant, so it was a situation whereby everyone is against you. At that point, I was really scared. So I went to stay with my friend for some months. Then I met with my brother who told me to come back home. When my family finally accepted the situation, they said, 'The fact remains that you are our daughter and our sister'. On November 13th, I gave birth to Kevo. I was seventeen.

Despite the commonness of teenage pregnancies in urban low-income neighbourhoods, the social stigma associated with teenage pregnancy was, as Eliza described, a kind of 'curse'. Her father and brothers had stopped speaking to her for a time, and the mothers of her friends forbade them to socialize with Eliza for fear that she was a bad influence. But Eliza reflected on this period as both a time of social exclusion but also an affirmation of her independence. She recollects her eldest brother being the central source of encouragement in this regard, insisting that, 'We'll support you, just give birth and have the baby, we'll help you take care of the baby. Then you'll go back to school'.

After Kevo was born, Eliza took a hairdressing course. Hair salons were one of the most common local businesses in the slums and most women paid up to 1000 KES (approximately 10 USD) to get their hair braided, equivalent to nearly a third of one month's rent in Huruma. But Eliza did not like hairdressing, and even less being bossed around with little prospect for advancement. She preferred to work independently, and following her hairdressing course started experimenting with various modes of self-employment. Working in one place for two to three months, she would use her saved earnings as seed capital for a small business idea, such as selling rice in her neighbourhood. Eliza travelled long distances once every two weeks to purchase rice at wholesale prices, and then sell a kilogram of rice through door-to-door sales to people she knew. She made a small profit on sales and saved on overhead costs, and her customers saved money by paying less than the local retail price. But as with any business in the hustle economy, there were inevitable blockades even in seemingly lucrative and well-oiled business models, which required quick adaptation and improvisation. When the price of rice went up, Eliza gave up the rice distribution business, and turned to her male peers and childhood friends, who by then had all finished or dropped out of secondary school. Most of them were engaged in waste work, which had become a common entry point into the informal economy in slums all over the city, as a result of ever-growing demographic growth with an ill-equipped municipality to service low-income areas.

This was in 2005, and very few women were doing garbage collection at the time. Eliza spoke about being the only woman 'doing this kind of job' in a way that highlighted the physical demands of the work and the pejorative associations with handling garbage. But her recollection of this work also emphasized the pride of being part of a largely youth-based entrepreneurial economy, and doing something that the local government had failed to provide: deliver basic waste management to a residential community. On the one hand Eliza admitted to being desperate enough to do almost any kind of work if it would support her son and the rest of her family; on the other hand like many other youth who had grown up in the 'ghetto', there was a deliberate choice in doing 'dirty work' on their own terms over 'donkey's work' for an exploitative or abusive boss in the industrial area of the city. Indeed working for someone else, for this generation of Kenyan youth, recalled highly uneven power relations and exploitative practices between employer and employee. Working for yourself, and amongst fellow peers, anchored in friendship-based youth groups, was integral to the collective identity of many young people living in the slums who claimed to enjoy the relative freedom of circumventing both formal state and formal market channels of service provision, employment, and control.

In addition to her income-generating work, Eliza was also a peer educator for young girls in her neighbourhood, something she had become involved with following Kevo's birth in pursuit of a support network for young mothers in the slums. It was this network, she claimed, that helped her complete secondary school in 2006.

> When I got that training, I learned that even if you are pregnant, even if you have given birth as a young girl, that does not mean that it is the end of everything. When I finished my Form Four exams, after that, people were proud of me. I started seeing some of the girls who were stressed with life saying, 'I've become pregnant. I have a child. Look at my family situation, what will I do? My only option is to go for early marriage'. That's where I would say, 'No, you have other options. You are now 18 years old. If you get married now, where do you think you will be in five years? Why can't you change your life and change your community?'

Eliza informally mentored numerous young women who experienced teenage pregnancy to think about their options beyond early marriage. As she put it, 'in the ghetto marriage is often an issue of peer pressure. You feel as if you are pushed against the wall'. Eliza's motto was that being single enabled a woman in the ghetto to think about her future. For her, that meant thinking about what it would take for her son to realize his potential, and to eventually attain the status of manhood according to her cultural tradition.

> For me personally, getting married is not an issue but what's an issue is having my little boy ask me, 'Where is my father?' You see, in the slums, if you get married and you already have a child, that family won't see your child as part of the family. If I got married my son wouldn't feel accepted into the family. And you know Kenyan men, even if you have a boyfriend, he'll not take care of your baby. So it's up to you as a mother to do it. First and foremost the single mothers will think about their child. The relationship with men comes second. But the bigger issue for me is one day having my son be able to say, 'There is my land'. Having some property for him. That is what I'm thinking instead of getting married. If I can provide him with a good situation, having some land where he can go, he will never ask you, 'Where is my dad?' or things like that, because I will have given him something of his own.

Here is where Eliza's story underscored the dialectics of identity for female youth living in the city. For her it would be paramount for Kevo to one day have land of his own. In her tribe, this was only possible through patrilineal heritage, where land was traditionally passed on from father to son. But without a father, a 'bastard son' would never be given land by a father figure. Any new man Eliza would let into her life, she explained, would never accept, let alone adopt, Kevo as his own son. Eliza's goal, what drove her everyday ambitions, was the notion that she could one day afford to buy a plot of land for her son, preferably somewhere rural. To have a claim to some land 'up-country' was an important part of one's narrative because even for youth who were born and raised in the city, they were (their kin, their tribe, their family farm) nevertheless from *somewhere* up-country. In this, she abided to traditions around gender and land inheritance. The means to this end would however be achieved by defying gendered norms of behaviour and practice.

In 2008, Eliza was one of the few dozen waste workers in Nairobi slums who became involved with a social enterprise called Community Cleaning Services (CCS). CCS was a Kenyan micro-franchise, sponsored by an American company committed to improving sanitation poverty through grassroots partnerships with low-income entrepreneurs who would offer cleaning services focused on shared public and community toilet facilities (Thieme, 2015). Eliza quickly went from being a team member of one of the CCS Mobile Cleaning Teams (MCTS), to a team lead and mentor, to a salaried CCS Professional responsible for training and following up with different CCS teams across Nairobi and reporting back to the Central CCS Management team. By time I met Eliza in 2009, she had become the sole breadwinner for her family, with her parents and siblings each out of work, and her six-year-old son in primary school.

'If they know how much you make, everyone will come over for lunch'

In each neighbourhood within Nairobi slums, there were informally marked territories known to youth groups as 'the base', where domestic, social, and business spaces often overlapped. The base of youth groups served an important geographical and social function. For instance, one of the youth groups Eliza frequently visited had a base situated at the bottom of the first hill one hundred metres from the *Matatu* stage on the main road linking Mathare Valley to the Central Business District. It was next to a public toilet they had built and managed since 2007. The guys could be found seated under a small wooden shed, where they had multiple angles of visibility. They could see who was coming from the stage; what the elders were doing on the other side of the footpath (often playing a local version of checkers and discussing local politics); they could collect payment for the toilet and water point both metres away from where they sat; they could see their urban farming corner to the left, their office to the right, and have a perfect view across the Mathare Valley. The base enabled peer groups to combine moments of loitering and killing time with a constant low-level flow of income generation that asserted their place and legitimacy in the public space of the neighbourhood. Eliza's street credibility involved having the license to kill time with these youth groups whose everyday practices blurred the lines between work, leisure, and idleness.

Although the base comprised an informally marked territory in the neighbourhood for most groups, there was always a kind of headquarters or sitting corner where the group assembled, greeted, schemed, hung out, bantered, began, or ended their working ventures. The base was a meeting point, where you could be found, where you belonged, and where you were known. It informed street credibility and relationships to the base had everything to do with one's visibility. When it came to garbage collection and servicing the plots of other tenants, youth asserted, 'You do not go beyond your base'. This relationship between economic activity and territorial zoning related both conceptually and practically to codes of youth group culture, reflecting a key characteristic of the informal waste economy: the small and fragmented scale of these otherwise neighbouring enterprises.

Contrary to most conceptions of entrepreneurship informed by neo-classical economic theories, the aspirations of these youth hustlers did not include maximizing growth. Eliza explained the potential negative effects of doing 'too well',

> Most youth don't want to admit how much they earn – even to each other. When you seem like you make more than your neighbour, you can get into trouble. And if you do well and everyone knows it, then you'll find you have lots of friends coming over your place right before lunch time!

Youth's individual strategic discretion regarding income was tied to the *mtaa* (hood) logic of solidarity inherent in everyday struggle. From the hustler's perspective, being part of a collective offered the benefit of risk reduction by diversifying ownership and reducing the traceability of any individual's gain from the entity. The difficulty of group membership, however, transpired for aspiring entrepreneurs setting themselves apart from group activities to pursue their individual ambitions while retaining a sense of belonging to the *mtaa*, their base, and their cohort.

In Huruma, Eliza was a self-proclaimed hustler, one of the guys, and part of the city's underworld. But part of maintaining her *hustler* persona *vis à vis* her peers within the *mtaa* required concealing to an extent that she was 'doing well'. Eliza had spent years struggling as a single mother pursuing various kinds of informal work while trying to finish school and manage her family obligations. When she started garbage collection work in 2005–2006, her entrepreneurial and resourceful capabilities coupled with ambitions to provide for her son as a single mother accelerated her productivity compared to her peers. It was this motivation that also fast-tracked her trajectory from member of a CCS team, to lead entrepreneur with multiple teams in Huruma, until she was finally hired by CCS as a full-time employee (Figure 7.3).

Her drive to prove herself within a male-dominated macho work culture defied commonplace outcomes of so many young single mothers in the hood, but her place within the traditional youth-elder hierarchies also meant that there would be no foreseeable end to the obligation to provide for the rest of her family with whom she still lived: her parents and two older unemployed brothers. Once the pride of being the main breadwinner subsided, she realized that her doing well gave her brothers little incentive to actually find work, and perpetuated a cycle of deference to her parents who expected her to share her earnings with the whole family as part of the reciprocal exchange involved following years of childcare support. As Collins et al. (2009, p. 2) state in their reflection on the vicissitudes of

Figure 7.3 Eliza with the boys, outside a public toilet at the end of a cleaning job.

Source: Kennedy Kariuki, 2010.

the informal economy, 'you make more on some days, less on others, and often get no income at all. Moreover, the state offers limited help, and, when it does, the quality of assistance is apt to be low. Your greatest source of support is your family and community, though you'll most often have to rely on your own devices'.

By early 2010, Eliza had saved up enough money to move out of her childhood home in Huruma, encouraged by a family friend to move far from Mathare area to assert and protect her financial and personal space. Six months later, she sent her seven-year-old son to boarding school, which was revered in Kenya as a tremendous education opportunity for any child, especially for anyone living in the slums where the quality and resources for formal education were scant, and where most mothers complained that the temptations for young men in the slums to get involved in drugs, street crime, and other kinds of trouble were constant. In theory, this 23-year-old woman had 'made it'. But when she moved away to a low-middle-income estate in the northeastern part of Nairobi, far away from everything and everyone she knew, Eliza admitted to feeling lonely and detached. During a follow-up visit in 2011, as we walked to her place one evening, I asked her if this neighbourhood was starting to feel like home. She said no, and as she often did, put it metaphorically: 'This estate is my bedroom. But the *mtaa* in Huruma is still my sitting-room'. Her work and social ties remained in Huruma. It is where she spent her Sundays, where she went to get her hair braided, where she hung about with the guys at their respective bases in between jobs.

Negotiating worlds within and between lives in the hood

The 'hustle' described in this chapter shows that the hood is not merely a temporary 'stop gap' for the emerging middle classes of 21st-century post-colonial cities, but rather an urban form in its own right, where social and economic spaces are being shaped by a generation of youth caught between life worlds and contrasting imaginaries. It has produced a generation of youth whose entanglements with everyday struggle shape their perceptions of the future, but the small nodes of external support, encouragement, and nudges to keep aspiring are both critical and fragile.

Since 2011, many notable events have marked Eliza's circumstances. In 2012, her family's farm up-country tragically burned down one night. The following day, her entire savings (almost 2,000 USD) went towards emergency expenditures. She had been 200 USD away from buying Kevo a plot of land she had had her eye on since 2008. In 2013, CCS had to close down, and Eliza lost her job. She immediately started another business, selling non-perishables in the now-growing neighbourhood where she lived, and she continues to work on sanitation-related projects across Nairobi, trying to revive CCS in some form. In 2014 she gave birth to her second child, telling me with a mischievous giggle over the phone that 'Kevo needed a baby sister'. She is still single; Kevo is still in boarding school and is adamant that he wants to one day become a doctor.

Eliza's experience illustrates the relative social exclusion that resulted from 'doing *too* well', conflicting with the strong sense of belonging that resided in youth who hustled within their *base* for the majority of their formative years. As Eliza defiantly claimed during my last visit, 'It's important to leave the *mtaa* and your base at some point, because it's not everyone who will be happy with your progress'. Yet, despite residing elsewhere, affording higher rent and upgrading living standards, Eliza and others like her continued their ties to the hood and the hustle economy. Being from the *mtaa* equated to a kind of badge, giving way to reverse prejudice towards young people in the city who romanticized and faked being from the hood. Where you eventually 'stayed' mattered less than where you had grown up, and where you went to socialize. Amongst those who had grown up in the hood, who worked where they lived, and for whom, 'in the ghetto, business is social', there was a perceived limit to growth and a cost to doing too well that was inextricably tied to one's street credibility as a hustler.

Eliza's life history shows how identity and sense of belonging were contingent on the protracted liminality of hustling, and how she negotiated the social implications of 'doing too well'. She chose to pursue her aspirations combining social work, activism, entrepreneurship, and further education, but has had to navigate the micro-politics of the hood amongst youth who know no other life world. Strategic discretion about income, returning each afternoon to loiter alongside the jobless corners of her peers, or combining formal education and 'advancement' with a loyalty for learning the streets, her story relays the complicated dynamic of a young single mother navigating a post-colonial city, where the contradictions of modernity, makeshift urbanism, and alternative politics of struggle that evade traditional avenues of political resistance and mobilization are redefining the terms of urban dwelling on the margins.

Eliza epitomizes the paradox between youth disaffection with existing realities of urban injustice and the commitment to stay fixed within the very neighbourhoods that exemplify the injustices associated with slum urbanism. Somehow, losing her savings and her stable CCS job never seemed to pose catastrophic results for Eliza. She managed, at every hurdle, to tap into 'zig zag' entrepreneurialism and hustle her way to the next opportunity. What seemed more difficult for Eliza was balancing her periods of sustained financial success and her place as a fellow hustler in the hood. Caring to maintain that balance kept Eliza and others like her in a kind of middle zone, where hustling was partly aspirational and filled with hopes to fight the traps of poverty, but where losing one's identity as a hustler came at a tremendous social cost. Her story depicts the paradoxical subjectivity of young people for whom growing up in challenging neighbourhoods can require negotiating the fine line between aspiring to get out of the quicksand of urban poverty, especially for young women, while choosing to remain rooted in the struggle of urban life in the hood. That struggle, embodied in everyday *hustle*, has become integral to the street credibility, belonging, and solidarity of young people growing up the 'ghettos' of in this African city.

References

Borden, I., 2001. *Skateboarding, space and the city: architecture and the body*. London: Bloomsbury Publishing.

Collins, D., Morduch, J., Rutherford, S., and Ruthven, O. 2009. *Portfolios of the poor: how the world's poor live on $2 a day*. Princeton: Princeton University Press.

Comaroff, J., and Comaroff, J., 2005. Reflections on youth: from the past to the postcolony. In A. Honwana, and F. DeBoeck, eds., *Makers and breakers: children and youth in post-colonial Africa*. Oxford: James Currey, pp. 19–31.

Comaroff, J., and Comaroff, J., 2012. *Theory from the south: or, how Euro-America is evolving toward Africa*. Colorado: Paradigm Publishers.

Chiuri, W., 1978. *Mathare valley: a squatter settlement in Nairobi: a case study*. Nairobi: UNICEF.

De Boeck, F. and Honwana, A., 2005. Introduction: children and youth in Africa. In: A. Honwana and F. DeBoeck, eds., *Makers and breakers: children and youth in postcolonial Africa*. Oxford: James Currey.

Diouf, M., 2003. Engaging postcolonial cultures: African youth and public space. *African Studies Review*, 49(1), pp. 1–12.

Dyson, J., 2008. Saka: growing up in the Indian Himalayas. In C. Jeffrey and J. Dyson, eds., *Telling young lives: portraits of global youth*, Philadelphia: Temple University Press.

Etherton, D., 1976. *Mathare Valley: a case of uncontrolled settlement in Nairobi*. Nairobi: University of Nairobi, Housing Research and Development Unit.

Fredericks, R., 2014. 'The old man is dead': Hip hop and the arts of citizenship of Senegalese youth. *Antipode*, 46(1), pp. 130–148.

Gidwani, V., 2001. The cultural logic of work: explaining labour deployment and piece-rate contracts in Matar Taluka (Gujarat) India, Part I and Part II. *Journal of Development Studies*, 38(2), pp. 57–108.

Hart, K., 1973. Informal income opportunities and urban employment in Ghana. *Journal of Modern African Studies*, 11(1), pp. 61–89.

Huchzermeyer, M., 2011. *Tenement cities: from 19th century Berlin to 21st century Nairobi.* London: Africa World Press.

Jauregui, B., 2014. Provisional agency in India: *Jugaad* and legitimation of corruption. *American Ethnologist*, 41(1), pp. 76–91.

Jeffrey, C., 2012. Geographies of children and youth II: Global youth agency. *Progress in Human Geography*, 36(2), pp. 245–253.

Jeffrey, C., and Dyson, J., 2013. Zig Zag capitalism: youth entrepreneurship in the contemporary global south. *Geoforum*, 49, pp. R1–R3.

Jeffrey, C., and Dyson, J., eds., 2008. *Telling young lives: portraits of global youth.* Philadelphia: Temple University Press.

Jones, G.A., 2012. Book review: hang about: young people's frustrations at the state of progress. *Dialogues in Human Geography*, 2(1), pp. 101–104.

King, K., 1996. *Jua Kali Kenya: change and development in an informal economy 1970–1995.* London: James Currey Ltd.

Mabala, R., 2011. Youth and 'The Hood': livelihoods and neighbourhoods. *Environment and Urbanization*, 23(1), pp. 157–181.

McFarlane, C., 2011. *Learning the city: knowledge and translocal assemblage.* Oxford: John Wiley & Sons.

Myers, G., 2005. *Disposable cities: garbage, governance and sustainable development in urban Africa.* Burlington, VT: Ashgate Publishing Company.

Myers, G., 2011. *African cities: alternative visions of urban theory and practice.* London: Zed.

Newell, S., 2012. *The modernity bluff: crime, consumption, and citizenship in Côte d'Ivoire.* Chicago: Chicago University Press.

Potts, D., 2007. The state and the informal in Sub-Saharan African urban economies: revisiting debates on dualism. Working Paper No. 18, Crisis States Working Papers Series No. 2.

Richardson, C. and Skott-Myhre, H.A., eds., 2012. *The habitus of the hood.* Chicago: University of Chicago Press.

Robinson, J. and Parnell, S., 2012. (Re)theorising cities from the global south: looking beyond neoliberalism. *Urban Geography*, 33(4), pp. 593–617.

Simone, A., 2010. *City life from Jakarta to Dakar.* New York: Routledge.

Sundaram, R., 2010. *Pirate modernity: Delhi's media urbanism.* New York: Routledge.

Thieme, T., 2010. Youth, waste and work in Mathare: whose business and whose politics? *Environment and Urbanization*, 22(2), pp. 333–352.

Thieme, T., 2013. The 'hustle' amongst youth entrepreneurs in Mathare's informal waste economy. *Journal of Eastern African Studies*, 7(3), pp. 389–412.

Thieme, T., 2015. Turning hustlers into entrepreneurs, and social needs into market demands: corporate-community encounters in Nairobi, Kenya. *Geoforum*, 59, pp. 228–239.

Vigh, H., 2006. *Navigating terrains of war: youth and soldiering in Guinea-Bissau.* New York: Berghahn.

Waage, T., 2006. Coping with unpredictability: 'Preparing for life' in Ngoundéré, Cameroon. In C. Christiansen, H. Vigh, and M. Utas, eds., *Navigating Youth, Generating Adulthood: Social Becoming in an African Context.* Uppsala: Nordic Africana Institute, pp. 61–87.

Wacquant, L., 2008. *Urban outcasts: a comparative sociology of advanced marginality*. Cambridge: Polity Press.

Weiss, B., 2009. *Street dreams and hip hop barbershops: global fantasy in urban Tanzania*. Bloomington: Indiana University Press.

Wilson, W., 1987. *The truly disadvantaged: the inner city, the underclass and public policy*. Chicago: University of Chicago Press.

8 From nomads to squatters

Towards a deterritorialization of Roma exceptionalism through assemblage thinking

Gaja Maestri[1]

Introduction

The alleged specificity of the Roma people has justified for a long time their control, stigmatization, and discrimination through means of spatial management that do not apply to other groups. First seen as vagrants, then as untrustworthy racial group, and finally as nomads: all of these representations have underpinned a set of exceptional policies for Roma only, like the adoption of halting sites in different European countries, ID cards for nomads in France, or EU policies for the integration of the Roma. This exceptionalism of the Roma people has also produced, and at the same time is reinforced by, a recent professionalization of pro-Roma NGOs and advocacy associations, as well as expert and academic knowledge production on the Roma as specific subject. This chapter interrogates the scholarly research on discourses and policies targeting the Roma, showing how it is possible to produce a knowledge that does not buttress their representation as exceptional category. I suggest turning to assemblage thinking to release the study of policies on the Roma from the risk of reproducing the Roma as separate subject.

The chapter consists of four main sections. In the first section I discuss the emergence of the Roma as a specific research and policy object, and the risk of essentialization posed by the discipline of Romani studies. In the second section I introduce assemblage thinking in order to understand the policies aimed at the Roma without, however, falling into the trap of Roma exceptionalism. The third and fourth sections of this chapter present the housing situation of the Roma in Rome and the specific case of the clearance of an informal settlement. The main aim is to illustrate how approaching the policies on the Roma from an assemblage angle can encompass disruptive practices that enable us not to reproduce the Roma as a fixed and exceptional group. The analysis draws on primary sources and data collected during my fieldwork in Rome from September to December 2013, during which I explored different forms of resistance to the spatial segregation of the Roma people.

The trap of Roma exceptionalism: discourses and policies

The last few decades have witnessed increased levels of attention towards the Roma people. In the early 1990s the EU, with the support of the Organization for Security and Cooperation in Europe (OSCE) and the Council of Europe (CoE), has started developing a minority and human rights policy framework to tackle the difficult conditions of the Roma (Sigona and Trehan, 2009). In 2005 European governments also joined the project Decade of Roma Inclusion, with the endorsement of the Open Society Foundation and the World Bank, whose aim was to eradicate Roma discrimination (Guy, 2009). New academic projects have been launched with the support of the European Union, like the European Academic Network on Romani Studies started in 2011 to foster research on issues regarding the Roma, and the MigRom project funded by the European Research Council.

Although these initiatives aim to overcome the marginalisation experienced by the Roma in Europe, and are an attempt to promote equality, they also contribute to the constitution of the Roma as an exceptional subject: homogeneous and separate from other marginalized groups. Furthermore, it is difficult to define who the Roma are. The CoE defines the Roma as a European minority, composed by two main groups: the Roma, mainly speaking Romani language, and the Travellers, conducting an itinerant lifestyle but not Romani speaking (Council of Europe, 2012). The European Roma Rights Centre (ERRC), in the call for applications for the 2013 Roma Rights Summer School, expands the definition of the CoE to 'other groups related to Roma and groups who are perceived by the majority populations as Roma'. As these definitions actually reveal, Roma and Travellers work as labels for many different groups, which are nonetheless lumped together. At an academic level, Willems (1997) and Tremlett (2009) argue that the discipline of Romani studies has been partly responsible for the essentialization of the Roma identity. Indeed, in this field the Roma have been defined either in terms of ethnicity and social class, or in terms of the changing definitions given by the majority population. In different ways both these conceptualizations portray the Roma as a homogeneous and exceptional group.

This essentialization of the Roma minority, despite being often used strategically to promote the rights of the Roma people (Vermeersch, 2005; van Baar, 2011), has some shortcomings too (Vermeersch, 2012). In fact it can exacerbate their exclusion from national narratives, the dismissal of national-specific forms of discrimination faced by the Roma, and it can result in the adoption of *ad hoc* policies which may reinforce their perception as an exceptional category (Vermeersch, 2012). This can also lead to a depoliticization of local policies on the Roma, as they come to be perceived predominantly as a European problem (van Baar, 2011), and also in the silencing of the Roma voice in favour of NGOs discourse (Trehan, 2009).

Although there is academic work challenging the exceptionalism of the Roma category,[2] scholars studying the Roma still rarely discuss their intersection with other categories, which can potentially lead to justify 'the reasons for exclusion and discrimination, and to provide more solid arguments supporting favour of markedly anti-gypsy initiatives' (Vitale and Claps, 2011, p. 251). This aspect is partly due to the presence of specific policies and discourses targeting the Roma that do not apply to other groups and for which there are also specific ways of analysis, as well as specific strategies of resistance. For instance, the discourse on nomadism and the policy of halting sites and evictions target only the Romani population, hence emphasizing the Roma as exceptional subjects. As a consequence, research focusing on these policies risks in turn reproducing a limited understanding of who the Roma are. How is it possible, then, to develop theories and research practices that, by starting from the analysis of policies targeting only the Roma, do not result in reproducing their position as exceptional category? Assemblage thinking can help tackle this conundrum.

Beyond Roma exceptionalism through assemblage thinking

The concept of assemblage refers to a composition of heterogeneous elements in 'some form of socio-spatial formation' (Anderson and McFarlane, 2011, p. 124). Deleuze and Guattari (1987) describe assemblage as a processual intermingling of different segments on two main axes. On the horizontal axis there are the segments of content and expression, while on the vertical one there are different degrees of territorialization (i.e., becoming stabilized) and deterritorialization (i.e., being carried away). The two horizontal segments of corporeal and incorporeal transformations are independent while at the same time intervening with each other. Understanding the relation between incorporeal and corporeal elements in terms of assemblage frees the reading of social formations of any linearity and brings the superlinear modes of assembling that Deleuze and Guattari define as 'rhizomatic' (Deleuze and Guattari, 1987) into the analysis. To understand a policy as assemblage underscores the process of coming together and falling apart of different elements without conceiving of them as fixed entities. More specifically, seeing policies as assemblages focuses attention on the different kinds of relationship that constitute them, as well as accounting for instability and change (McCann, 2011). Moreover, assemblage thinking can provide new insights into the analysis of urban phenomena and 'a deeper understanding of what policy actually does' (McFarlane, 2011b, p. 385), by highlighting the practices that constitute it beyond the explicit intent of policies (McCann and Ward, 2012).

With regard to this chapter, assemblage thinking can offer two main contributions: firstly, it draws attention to the active role of materialities in the constitution and contestation of inequalities (McFarlane, 2011a), and secondly, it permits a view of agency as an emergent aspect of an assemblage, therefore focusing on disruption (McCann, 2011). Although the English term *assemblage* might give the impression of a static arrangement (Gregory et al., 2009), the French term *agencement* suggests a more dynamic idea of an 'agency making entity',[3] whereby the arrangements constitute the capacity of acting in different ways. To conceive of policies as 'agency-making-entity' allows us to understand how they restrict agency, but also how they create conditions for practices of elusion and breakage (Deleuze, 1992). Indeed, Deleuze (2004) defines practice as the 'mutation point' and 'a point of permanent revolution' (Deleuze, 2004, p. 191) of a structure, or assemblage, hence enabling us to account for the potentiality of change from *within* the assemblage itself, and without seeing change as a mere result of supposedly external factors. While, for instance, Bourdieu stresses the role of habitus as relentlessly reproducing the continuity of certain structures (Bourdieu, 1990), in a Deleuzian sense practices are also disturbing and can lead an assemblage to fall apart, bringing about new intersections with other assemblages. Assemblages are therefore constantly producing the conditions for their crumbling, not only through expressive articulations, but also through material configurations. In the rest of the chapter, after outlining the context of the segregation and governance of the Roma in Rome, I will show how looking at the policy of eviction from an assemblage perspective can shed light on the role of space and movement in the emergence of disruptive practices.

The residential segregation of the Roma in Rome

It is estimated that there are 170,000–180,000 Roma and Sinti in Italy (UNAR, 2012), and that almost one-third of them experience severe housing deprivation, living in either official or informal settlements (Sigona, 2007) (Figure 8.1). Recent figures show that Rome is the Italian city with the highest concentration of Roma living in settlements (Dalla Zuanna, 2013). In 2012 the City of Rome counted a total of 11,021 Roma: 6,097 in 19 official settlements, and the remaining others in about 130 informal settlements (depending on the figures, they are between 80 and 200) (Dalla Zuanna, 2013). Of the 19 official Roma settlements, eight (called 'equipped villages') are fully equipped with Portakabins and basic facilities, while the remaining 11 (named 'tolerated camps'), although being officially acknowledged, are not directly run by the City of Rome. Despite the description of the Italian context as an 'anomaly' (Sassoon, 2013) and the 'urban narratives of divided and underdeveloped city' (Cremaschi, 2014, p. 67) characterizing the city of Rome, the Italian capital is actually situated within the global tendency of proliferation of camps and ghettos (Agier, 2011, 2014), and of the use of informality as governmental tool (Roy and AlSayyad, 2004; McFarlane and Waibel, 2012). People living in these settlements often experience residential segregation and severe housing deprivation, as the settlements often lack basic facilities, such as access to drinking water and electricity, and are mainly concentrated in urban fringe, and in non-residential and isolated areas far from services and public transport.

The first official settlements were created at the beginning of the 1990s in order to relocate the increasing number of Roma living in 'slums'. Most had arrived in Rome as asylum seekers during the Yugoslav Wars. The official settlements policy has since its inception been marked by an ambiguous intent, between hostility and hospitality, between humanitarian help and control (Stasolla, 2012). This approach reflects attitudes towards the Roma, seen as both in need of help and threatening. In Italy, as in the rest of Western Europe, the otherness and exceptionalism of the Roma have been mainly framed in terms of their supposed nomadism, that has characterized the Roma as different from the majority sedentary population (Simhandl, 2009; Vitale and Claps, 2011; Picker and Roccheggiani, 2013). The presence of a dominant 'sedentary metaphysics' (as defined by Malkki, 1992), whereby the relationship between culture and place is naturalized as fixed, leads

Figure 8.1 The settlement in Cesarina Street.
Source: Gaja Maestri, 2013.

to the stigmatization of uprooted and displaced people, such as the refugees, and in general of those who do not fall into the sedentary imaginary – therefore nomads too (Kofman, 1995). Furthermore, nomadism can be '*perceived* as disorderly and threatening' (Sibley, 1981, p. 38). Thus, nomadism as a discourse, regardless of whether practiced or not – today it is estimated that about 15% of Roma and Sinti in Italy have an itinerant lifestyle (Sigona, 2005) – is an important factor underpinning the housing exclusion experienced by the Roma in Rome. For instance, Sigona (2003, 2005) maintains that labelling practices defining the Roma as nomads are used as a justification not to provide public housing for the Romani population, who allegedly do not want to live in houses. Similarly, Hepworth (2012) argues that, despite Italian Roma's formal entitlement to the same housing policies offered to other Italian citizens, policymakers actually fail to promote housing inclusion by claiming to protect Roma nomadism as a cultural feature.

This situation was exacerbated during the last 15 years when local and national political debates on Roma informal settlements came to be prominently framed in security terms (Vitale, 2009). The former centre-left mayor Walter Veltroni (2001–2006; 2006–2008) increased the number of evictions and the construction of Roma official settlements in the periphery of the city (Stasolla, 2012). Between 2003 and 2007 it is estimated that almost 4,000 Roma were evicted and displaced within the city of Rome (Bonadonna, 2011). Today the official settlements are really overcrowded and characterized by unhealthy and segregating conditions.

Thus the City of Rome tends to evict Roma from informal settlements but rarely relocates the families to temporary so-called Roma reception centres, and most of the time only offers women and children relocation to shelters. As a consequence, many women refuse the offer in order to avoid separation from their husbands, with the result that the families are just left homeless, and scatter. Furthermore, most of the Roma experiencing severe housing deprivation are practically denied access to social housing policies. As shown by an Amnesty International report (Amnesty International, 2013), the Roma living in settlements often fail to gain eligibility for public housing because, for instance, they are not tenants or home-owners, therefore lack the correct status needed for a place on the social housing register. The Roma living in both official and informal settlements of Rome appear to be caught in a vicious circle whereby they are constantly nomadized through evictions, with a widespread access to Roma reception centres, limited access to official settlements, and almost no access to public housing.

This process of housing exclusion is reinforced by a set of administrative practices and policies enforcing the discourse of nomadism. For instance, in 2008 the Italian government adopted a Nomads Emergency Decree (Consiglio dei Ministri, 2008) that targeted Romani settlements, increasing their control, surveillance, and evictions. In the last 20 years the city of Rome has adopted three so-called Nomads Plans that consist of a set of measures targeting Romani communities experiencing severe housing deprivation, and whose precarious housing conditions are seen as part of their alleged nomadic lifestyle. In 1996 the City of Rome has set up a Nomads Office that is in charge of Roma official settlements and emergency reception centres. The Nomads Office is part of the Department of Promotion of Health and Social Works and it is different from the Department of Work, Housing, and Housing Emergency, which is responsible for social housing policies. This division reinforces the distinction of the Roma from other categories, whose housing needs are managed by a different department. The Nomads Office also cooperates with a special police unit, which is responsible for the monitoring of Roma settlements and the enforcement of clearance of Romani informal settlements. This police unit was called until 2011 Coordination of Interventions and Operations on Nomads (CION), and it has now become the Unit of Public Security and Emergency. By effecting what they name, these administrative labels, policies, practices, and knowledge regarding the Roma population contribute to their articulation as exceptional subjects. Indeed, this policy channel for Roma reproduces their alleged nomadism as an effect: It 'nomadizes' the Roma (van Baar, 2011) by making them constantly move from one informal settlement to another, and by accommodating them in caravans or Portakabins instead of houses. Given all of these constraints that relentlessly produce the Roma as a separate category through different discourses and material practices, the question is, how is it possible to analyze these situations without reproducing the Roma as exceptional subject? In the next section I investigate the case of the eviction from the Romani settlement Casilino 700 in 2009 in Rome. By employing the conceptual tool of assemblage I aim to show that it is possible to offer an account of this eviction that overcomes this apparent impasse.

From nomads-being to squatters-becoming

In November 2009 the residents of the informal Romani settlement Casilino 700 were evicted after a prolonged period of resistance. Since the early 1990s this informal settlement was located in the southeastern periphery of the City of Rome, in the VII Municipality where there were five other informal settlements. Casilino 700 (located in Casilina Street) was considered one of the biggest Romani settlements in Europe with more than 1,200 people estimated to be living there by the end of the 1990s (mainly Roma from Bosnia, Montenegro, Romania, and Macedonia) (Rossi, 2006). When the Roma migrants started squatting on this piece of land formerly used as an airport, they lived mainly in shacks and old caravans. The settlement lacked access to water, and the City of Rome provided only few toilets. Therefore hygiene, health, and safety were poor and this led to a couple of tragic accidents in which children lost their lives because of fires and toxic exhalation. Casilino 700 was controlled by the police and, because of its illegality and unhealthy conditions, residents were evicted for the first time in 2000. They were relocated to a new official settlement not far from Casilina Street – called Salviati 2 to distinguish it from another settlement already present in Salviati Street. Others were sent to another settlement in the southwestern periphery of the city called Tor De' Cenci, and then evicted again in 2012 (Rossi, 2006). Although there were a series of associations working in partnership with the local municipality to support settlement residents, the eviction could not be halted. After the first eviction a new informal settlement reemerged on the same site in 2008, mainly with Romani families from Southern Romania, but they were finally evicted in 2009. When the 360 residents of the new Casilino 700 received the eviction order in April 2009, the association Popica[4] that was working in the settlement managed to mobilize other associations and schools and to postpone the eviction. In June a second eviction order arrived and, on 12 November 2009 the police cleared the settlement. As a response 150 of the 360 evicted residents occupied an abandoned Heineken factory nearby, but they were eventually evicted from there too.

The evicted residents of Casilino 700 were offered temporary accommodation in the reception centre Cartiera (Figure 8.2), which was opened for that occasion in the northern part of the city. Almost 100 out of the 360 residents of Casilino 700 agreed to relocate to the Cartiera centre, which was located in an abandoned paper factory on a busy high street called Salaria Street and right next to a garbage disposal site (for this reason the air in the centre smelled really bad). The building was not built to house families, so at the beginning they slept together in the main hall. After the Associazione 21 Luglio[5] complained, the municipality provided sets of rooms, of about 4mq each, in order to give the families more privacy. The families spent most of their time in the backyard cooking, washing, and socializing because the building could become quite hot during summer and was not equipped for daily activities. During an informal conversation, the director and one of the centre managers argued that the Roma people were not suited to indoor life in houses, and were culturally used to living in the open air, therefore the Cartiera centre offered them an ideal environment for their cultural habits.

Figure 8.2 The main entrance of the Roma reception centre Cartiera in Rome, where the City of Rome offered temporary accommodation to the evicted residents of Casilino 700.

Source: Gaja Maestri, 2013.

By just focusing on the relocation of the Roma to the Cartiera centre we would, once again, include only the Roma in our analysis. However, adopting assemblage thinking as a theoretical tool to analyze the specific case of the eviction of Casilino 700 residents can help overcome this limitation for two main reasons. In the first instance, seeing the policy of eviction as an assemblage means rejecting a univocal understanding of eviction as simply the act of clearing sites and relocating, hence entailing its unexpected effects. Evictions are indeed functional to the nomadization of the Roma (van Baar, 2011) – as in the case of the relocation to the Cartiera centre – but they also enable forms of resistance, setting up the conditions for deterritorializing practices towards other assemblages. Secondly, approaching the Roma eviction policy as assemblage enables us to see the revolutionary potential of spaces and movements. An eviction is a performative act that produces the effect of marginalizing the Roma only through specific practices, that is, if the evicted Roma follow preestablished options, such as relocation to Roma reception centres or the formation of other informal settlements. But, what if the evicted Roma do not enact preestablished movements and an eviction is followed by unexpected practices? And, most importantly, what happens if we include these disruptions in the definitions that are given to the Roma?

After the eviction, the 150 residents of Casilino 700 who attempted to occupy the Heineken factory joined the political squat Metropoliz (for a classification of squatting, see Prujit, 2013). This latter had been occupied since May 2009 by the movement for housing rights Blocchi Precari Metropolitani (BPM).[6] Metropoliz (Figure 8.3) is the first squat (and the one with the largest Romani community) where Roma people cohabit with other homeless migrants and Italians. Initially the Roma families decided to start building their houses in a hangar in front of the main building of the factory. After one year some families decided to move right outside the hangar, and others decided to move into the main building with the rest of the occupants. Most of the Roma eventually moved into the main building following the clearance and destruction of the hangar in August 2012, while a small number left the squat. In Metropoliz there are different activities such as art workshops and a football team for the children living there, mainly organized by the members of the BPM and the people working and volunteering for Popica. Since 2009 there have been also other squats involving Roma groups set up by the BPM and the Resistenza Abitativa Metropolitana (RAM) in the eastern periphery of Rome, and by the group Action in the southern outskirts. In these cases the Roma started mobilizing together with other migrants and Italians as part of a new political subject not based on their ethnicity and exceptionality, but based on their shared socio-economic status of being excluded from the formal labour market in times of crisis (Maestri, 2014).

The Roma who joined Metropoliz enacted their eviction differently from those who moved to the Cartiera centre, like a rhizome that grows in unexpected multiple ways. While reterritorializing practices of evictions imply the relocation to a temporary Roma reception centre, or the setting up of other informal settlements (as in the case of the remaining 100 former residents of Casilino 700), the relocation to a squat organized by a movement for housing rights constitutes a move from nomads-being to squatters-becoming. The discourse on nomadism also hinges on the performative effects of evictions that nomadize the Roma: by performing the eviction differently, the Roma who joined Metropoliz managed to escape the frame of nomadism – and so did we escape the trap of the Roma exceptionalism by looking at this eviction differently. As a member of a pro-Roma local association puts it:

> They became squatters [. . .] who ask the municipality for a housing solution, not as foreigners or Roma, but as citizens. It doesn't take much! And then no one from the Nomads Office bothered them anymore [. . .] they have been considered another thing.

Although they did not claim this act was a way to challenge the discourse on nomadism, the Roma living in Metropoliz have actually unsettled the assemblage of policies based on it. Indeed, while as Roma they are offered a Roma-only solution, as squatters they are offered relocation to either public housing or other emergency housing solutions (for non-Roma). For example, Metropoliz has been included in the list of squatted buildings mentioned in resolution no. 206 (Comune di Roma, 2007). This resolution was approved in 2007 and provides that in the case of eviction 15% of available public housing should go to the occupants of squats for housing rights, therefore to the Roma of Metropoliz too.

Figure 8.3 Entrance of the main building of the Metropoliz squat.

Source: Gaja Maestri, 2013.

By practicing a movement to a space (a political squat) that falls outside the reach of the Roma policy assemblage, the Roma managed to escape the marginalizing effects of evictions, leading to a rhizomatic rearticulation of their marginalized – rather than marginal – position. As discussed in the previous sections, marginalization is a process emerging from different practices and discourses, which is nonetheless always open to continuous negotiation, contestation, and reassembling. For instance the Roma living in Metropoliz, and in general those who joined other squats for the fight for housing rights, are no longer subject to evictions carried out by the police Unit of Public Security and Emergency because they have entered a new policy assemblage and become a different political subject: squatters. This shows that policies and practices of control and management are based on fixed categories and that becoming something different from what one subject is supposed to be can trouble the functioning of a policy apparatus. As an interview with a police officer member of the Unit of Public Security and Emergency shows, the Roma living in occupied buildings squatted by social movements are not under the jurisdiction of this police unit anymore:

> We report everything to the department [of Promotion of Health and Social Works] so that they know what's going on [. . .] but usually these things are the responsibility of the riot police, the state police or the Carabinieri. . . .

Similarly another interviewee, a member of the Nomads Office in Rome, argued that while issues regarding Roma informal settlements are dealt with by the Nomads Office and social workers, squatted buildings are not their responsibility.

By 'studying through' (McCann and Ward, 2012) the very practices that produce nomadism as an effect, rather than looking at the policies towards the Roma as we know them, we could broaden the range of different policy assemblages in which the Roma are part of – such as the squatting assemblage. As Lancione (2013) argues, to change normative policies that are based on stigmatizing knowledge, the mechanisms that generate a policy assemblage should be deterritorialized by starting from assemblage itself.

Conclusion

This chapter has illustrated how, by conceiving of policies as assemblages, it is possible to avoid replicating normative assumptions. Unintentionally, and in many cases with critical purposes, research into policies on the Roma often contribute to the production of a knowledge on the Roma as exceptional subjects, separate from other categories and domains studied by different disciplines. Since the alleged exceptionalism of the Roma is one of the aspects that justify the policies targeting them, I have argued that there is a need for academic research that does not reproduce this separateness. Changing the way research on the Roma is done is important, as academics and researchers are crucial actors in the constitution of the knowledge on the Roma, hence also the power relations within which they are implicated. This can be done by trying to understand the process of the becoming of Roma subjectivity rather than their fixed identities, focusing on what goes beyond the definition of the Roma as we know them, and treating this difference as a constitutive part of the research object. The concept of assemblage has helped encompass other assemblages which the Roma have become part of, in order to understand their political subjectivity not as exceptional but as becoming other from what they are usually understood to be. For example, the case of Metropoliz as a site of intersection of the assemblage of Roma policies and of squatting: by following the disruptive practices of an eviction of residents from a Romani settlement, the analysis has included a non-Roma space, a political squat, where the Roma joined a new political subject based on the condition of housing exclusion experienced by a different set of categories in times of crisis. Are, then, the Roma really as exceptional as the discourses and policies on them make us believe? The answer is negative and can be found by approaching the policy of Roma evictions as assemblages composed by different elements. Starting from disruptive practices it is possible to produce an unexpected and troubling glimpse of categories that are taken for granted.

Notes

1 I would like to thank Angharad Closs Stephens, Joe Painter, Jennifer Bagelman, Madhumita Dutta, and Nicolò Palazzetti for their insightful and useful comments. Special thanks go to Michele Lancione for his help in improving the quality of this work. I am

also thankful to Lucy Smout Szablewska for proofreading the chapter. This work was supported by the Economic and Social Research Council [ES/I007296/1].

2 See, for instance, Okely (1983) who rejects the ethnic definition of the gypsies, Fassin (2010) who compares the stigmatization of the Roma to that of African-Americans and Muslims in Europe, and Picker and Roccheggiani (2013) who advocate more historical comparative research on the stigmatization of the Roma and the Jews.

3 This English translation of the French term *agencement* was suggested by Professor Rosi Braidotti in the lectures she held during the Utrecht Summer School course Critical Theory, Culture, and Citizenship (12–16 August 2013).

4 Popica is a non-profit organization working in Italy (in Rome) and in Romania on issues such as child protection, social exclusion, and homelessness. Popica refuses to work in official settlements because they do not agree with their segregating effects, and runs activities mainly with Romanian Roma communities living in informal settlements.

5 The Associazione 21 Luglio, i.e., Association 21 July, is a pro-Roma advocacy group founded in 2008. It is one of the most active associations which openly criticize the Roma housing policies adopted by the City of Rome, and often collaborates with Amnesty International and the European Roma Rights Centre (ERRC).

6 The group Blocchi Precari Metropolitani (i.e., Precarious Metropolitan Blocks) is part of a housing movement in Rome which aims to provide housing for homeless and displaced people and also denounces the role of real estate speculation in the gentrification processes. Two other groups that I mention later in the text, Action and Resistenza Abitativa Metropolitana (Metropolitan Housing Resistance in English), are also part of the housing movement.

References

Agier, M., 2011. From refuge the ghetto is born: contemporary figures of heterotopias. In: R. Hutchison and B.D. Haynes, eds., *The ghetto. Contemporary global issues and controversies*. Boulder: Westview Press, pp. 265–292.

Agier, M. ed., 2014. *Un monde de camps* [A world of camps]. Paris: La Découverte.

Amnesty International, 2013. *Double standards. Italy's housing policies discriminate against Roma.* London: Amnesty International Ltd.

Anderson, B. and McFarlane, C., 2011. Assemblage and geography. *Area*, 43(2), pp. 124–127.

Bonadonna, F., 2011. La segregazione amichevole per gli 'zingari' in Italia [The friendly segregation of gypsies in Italy]. *Alternative per il Socialismo*, 14(1), pp. 192–228.

Bourdieu, P., 1990. *The logic of practice*. Stanford, CA: Stanford University Press.

Comune di Roma, 2007. *'Patto per Roma sicura' del 18–05–2007* [Pact for a safe Rome]. Rome: Comune di Roma.

Consiglio dei Ministri, 2008. *Decreto del Presidente del Consiglio dei Ministri, 21 maggio 2008. Dichiarazione dello stato di emergenza in relazione agli insediamenti di comunita' nomadi nel territorio delle regioni Campania, Lazio e Lombardia.* [Decree of the President of the Council of Ministers, 21 May 2008. Declaration of state of emergency with regard to settlements of the nomadic communities in the territory of the regions Campania, Lazio and Lombardy].

Council of Europe, 2012. *Council of Europe descriptive glossary of terms relating to Roma issues* [online]. Available at: <http://a.cs.coe.int/team20/cahrom/documents/Glossary%20Roma%20EN%20version%2018%20May%202012.pdf> [Accessed 6 Jun. 2013].

Cremaschi, M., 2014. Worlding, worldly or ordinary? Repositioning Rome. In: A.G. Calafati, ed., *The changing Italian cities: emerging imbalances and conflicts,* GSSI Urban Studies – Working Papers 6. L'Aquila: Gran Sasso Science Institute, pp. 67–74.

Dalla Zuanna, G., 2013. *La popolazione in forte disagio abitativo in Italia. La condizione dei richiedenti asilo, dei rifugiati e dei Rom* [Populations experiencing severe housing deprivation. The condition of asylum seekers, refugees and Roma]. Cittalia – ANCI.

Deleuze, G., 1992. What is a dispositif? In: T.J. Armstrong, ed., *Michel Foucault philosopher*. Hertfordshire: Harvester Wheatsheaf, pp. 159–166.

Deleuze, G., 2004. How do we recognize structuralism? In: D. Lapoujade, ed., *Desert islands and other texts 1953–1974*. New York: Semiotext(e), pp. 170–192.

Deleuze, G. and Guattari, F., 1987. *A thousand plateaus*. Minneapolis: University of Minnesota Press.

Fassin, E., 2010. Pourquoi les Roms? [Why the Roma?]. *Mediapart*. Available at: <http://blogs.mediapart.fr/blog/eric-fassin/120910/pourquoi-les-roms> [Accessed 12 Jan. 2013].

Gregory, D., Johnston, R., Pratt, G., Watts, M. and Whatmore, S. eds., 2009. *The dictionary of human geography*. 5th ed. Oxford: Blackwell.

Guy, W., 2009. EU initiatives on Roma: limitations and ways forward. In: N. Sigona and N. Trehan, eds., *Romani politics in contemporary Europe: poverty, ethnic mobilization, and the neoliberal order*. Basingstoke: Palgrave Macmillan, pp. 23–50.

Hepworth, K., 2012. Abject citizens: Italian 'nomad emergencies' and the deportability of Romanian Roma. *Citizenship Studies*, 16(3–4), pp. 431–449.

Kofman, E., 1995. Citizenship for some but not for others: spaces of citizenship in contemporary Europe. *Political Geography*, 14(2), pp. 121–137.

Lancione, M., 2013. Homeless people and the city of abstract machines: Assemblage thinking and the performative approach to homelessness. *Area*, 45(3), pp. 358–364.

Maestri, G., 2014. The economic crisis as opportunity: how austerity generates new strategies and solidarities for negotiating Roma access to housing in Rome. *City: Analysis of Urban Trends, Culture, Theory, Policy, Action*, 18(6), pp. 808–823.

Malkki, L., 1992. National geographic: the rooting of peoples and the territorialization of national identity among scholars and refugees. *Cultural Anthropology*, 7(1), pp. 24–44.

McCann, E., 2011. Veritable inventions: cities, policies, & assemblage. *Area*, 43(2), pp. 143–147.

McCann, E. and Ward, K., 2012. Assembling urbanism: following policies & 'studying through' the sites & situations of policy-making. *Environment and Planning A*, 44(1), pp. 42–51.

McFarlane, C., 2011a. Assemblage and critical urbanism. *City: Analysis on Urban Trends, Culture, Theory, Policy, Action*, 15(2), pp. 204–224.

McFarlane, C., 2011b. On context. *City: Analysis on Urban Trends, Culture, Theory, Policy, Action*, 15(3–4), pp. 375–388.

McFarlane, C. and Waibel, M. eds., 2012. *Urban informalities. Reflections on the formal and informal*. Avebury: Ashgate.

Okely, J., 1983. *The traveller-gypsies*. Cambridge: Cambridge University Press.

Picker, G. and Roccheggiani, G., 2013. Abnormalising minorities. The state and expert knowledge addressing the Roma in Italy. *Identities: Global Studies in Culture and Power* [online]. Available at: <10.1080/1070289X.2013.854719>.

Prujit, H., 2013. Squatting in Europe. In: Squatting Europe Kollective, ed., *Squatting in Europe. Radical spaces, urban struggles*. Wivenhoe/New York/Port Watson: Minor Composition, pp. 17–60.

Rossi, M., 2006. Rom a Roma. Pratiche di integrazione e controllo: il campo di via Casilina 700 [Roma in Rome. Practices of integration and control: the settlement in 700 Casilina Street]. In: S. Misiani, P.P. Mudu, D. Pessina, M. Rossi and E. Sonnino, eds.,

La periferia perfetta. Migrazioni, istituzioni e relazioni etniche nell'area metropolitana romana. Milano: FrancoAngeli, pp. 221–243.
Roy, A. and AlSayyad, N., 2004. *Urban informality: transnational perspectives from the Middle East, Latin America, and South Asia*. Oxford: Lexington Books.
Sassoon, D., 2013. The Italian anomaly? *Comparative European Politics*, 11(3), pp. 280–295.
Sibley, D., 1981. *Outsiders in urban societies*. New York: St. Martin's Press.
Sigona, N., 2003. How can a 'nomad' be a refugee? Kosovo Roma and labelling policy in Italy. *Sociology*, 37(1), pp. 69–80.
Sigona, N., 2005. Locating 'The Gypsy problem'. The Roma in Italy: stereotyping, labelling and 'nomad camps'. *Journal of Ethnic and Migration Studies*, 31(4), pp. 741–756.
Sigona, N., 2007. Lo scandalo dell'alterità: rom e sinti in Italia [The scandal of alterity: Roma and Sinti in Italy]. In: S. Bragato and L. Menetto, eds., *E per patria una lingua segreta. Rom e sinti in provincia di Venezia*. Portogruaro: nuovadimensions, pp. 17–32.
Sigona, N. and Trehan, N., 2009. *Romani politics in contemporary Europe: poverty, ethnic mobilization, and the neoliberal order*. Basingstoke: Palgrave Macmillan.
Simhandl, K., 2009. Beyond boundaries? Comparing the construction of the political categories 'Gypsies' and 'Roma' before and after EU enlargement. In: N. Sigona and N. Trehan, eds., *Romani politics in contemporary Europe: poverty, ethnic mobilization, and the neoliberal order*. Basingstoke: Palgrave Macmillan, pp. 72–93.
Stasolla, C., 2012. *Sulla pelle dei rom* [On the skin of the Roma]. Roma: Edizioni Alegre.
Trehan, N., 2009. The Romani subaltern within neoliberal European civil society: NGOization of human rights and silent voices. In: N. Sigona and N. Trehan, eds., *Romani politics in contemporary Europe: poverty, ethnic mobilization, and the neoliberal order*. Basingstoke: Palgrave Macmillan, pp. 51–71.
Tremlett, A., 2009. Bringing hybridity to heterogeneity in Romani Studies. *Romani Studies*, 19(2), pp. 147–168.
UNAR, 2012. *National strategy for the inclusion of Roma, Sinti and Caminanti communities*. European Commission Communication no. 173/2011.
Van Baar, H., 2011. Europe's Romaphobia: problematization, securitization, nomadization. *Environment and Planning D: Society and Space*, 29, pp. 203–212.
Vermeersch, P., 2005. Marginality, advocacy, and the ambiguities of multiculturalism: notes on Romani activism in Central Europe. *Global Studies in Culture and Power*, 12, pp. 451–478.
Vermeersch, P., 2012. Reframing the Roma: EU initiatives and the politics of reinterpretation. *Journal of Ethnic and Migration Studies*, 38(8), pp. 1195–1212.
Vitale, T., 2009. Rom e sinti: i dilemmi dell'azione pubblica [Roma and Sinti: the dilemmas of public action]. *ERE Emilia-Romagna-Europa*, 2, pp. 94–104.
Vitale, T. and Claps, E., 2011. Not always the same old story: spatial segregation and feelings of dislike against Roma and Sinti in large cities and medium-size towns in Italy. In: M. Rovid and Mi. Stewart, eds., *Multi-disciplinary approaches to Romany studies*. Budapest: Central European University Press, pp. 228–253.
Willems, W., 1997. *In search of the true gypsy: from enlightenment to final solution*. London: Frank Cass Publishers.

9 The machine and the poet

A tale of how the subject goes into the field (and how it comes back)

Jean-Baptiste Lanne

This is Nairobi! [. . .] If there is a miracle in the idea of life it is this: that we are able to exist for a time, in defiance of chaos. [. . .] For us, life is about having a fluid disposition. Nairobi is a shot of whisky.

Binyavanga Wainaina, *Discovering Home*, 2003[1]

On 31 March 2012, Kiplangat and I are on our way to the wedding ceremony of Abeid and Zulaikha. I have been doing a fieldwork for four months in Kibera, one of the biggest slums in Nairobi (Kenya). I usually introduce myself as a geographer, and people usually laugh about my non-geographical object: festivity. In the field, I spend most of my time explaining why I am wandering around. Indeed festivity does not happen anywhere, anytime. One day, I met Kiplangat, a 30-year-old man from Olympic (one of the 15 villages composing Kibera), in front of his second-hand clothes shop. He told me about something that might interest me: a forthcoming wedding between a Nubian young man from Makina (another village), and a Luo[2] woman, recently converted to Islam. We could go together to the ceremony and stay for the party afterwards, he said. He offered to host me for the night. I could not miss this double opportunity: witnessing a traditional wedding – or what I thought it would be, with a cup of naiveté, I admit it – and having a little walk across Kibera by night.

A wedding in Makina: three recordings[3]

Please consider the three following recordings, extracted during and after the wedding.

Recording 1 – procession – around 6:00 p.m.

Just after the religious ceremony at Makina Mosque, all guests walk in a long and noisy procession across the narrow streets of Makina. I notice the rubbing of the Sunday clothes against the muddy walls of the houses. The temperature is incredibly hot (35°C? Maybe more?). Kiplangat and I choose to stay a few metres behind the procession. Question of temperature, and legitimacy: neither Kiplangat nor I are relatives of the couple's families. After a while, the bride Zulaikha, at the head of the procession, stops and enters her father's house. In the Nubian tradition, the new bride must stay three days and three nights in her former room. Mother, sisters, and female relatives keep watch over her. On the first day, other women from the procession are allowed to visit her and bring some presents. Kiplangat and I, amongst other men from the procession, are waiting in the courtyard in front of the house. Kind of a funny picture: this gathering of men, waiting, desiring to see what is going on inside, and respectfully keeping a distance. Some of them hang their arms to clotheslines (see Figure 9.1) in order to put up with a long and uncomfortable stand.

Standing at the doorstep, a woman notices me in the crowd, grasps my arm, and tells me I should come inside. She touches the camera tied to my wrist. I feel embarrassed, but at the same time, I really am eager to see Zulaikha dressed in her *gurbaba* (large colourful traditional Nubian dress), cautiously guarded by her little troop of Amazons. Kiplangat pushes me from behind to make me go. The woman insists: I should come inside and take some pictures. I am not the one who decides to come and cross the doorstep. Some people decide it for me: this woman, pulling my camera, and Kiplangat, pushing my shoulders. Both are laughing.

What happens inside? My presence does not seem to disturb anyone. More: the entire scene is – spatially, energetically – organized around me and my camera. Women are posing in front of me, placing themselves on both sides of the new

Figure 9.1 Men waiting under clotheslines in Makina, Kibera.

Source: Jean-Baptiste Lanne, 2012.

Figure 9.2 The bride Zulaikha in her room, surrounded by female relatives.

Source: Jean-Baptiste Lanne, 2012.

bride to fit in the frame (see Figure 9.2). No one asks about intrusion of a masculine presence. I am here: it is a fact. I press the button of the camera, again and again. *Me-and-my-camera*: a single entity, a single machine.

Recording 2 – dinner – around 8:00 p.m.

All guests are now entering Abeid's father's house for dinner. The house is divided in two different spaces: the front yard is converted into a kitchen (long tables, gas cylinders, huge bags of semolina) and is reserved to women, while men are talking, chewing *khat*, and dining in the only closed room of the house. Communications between these two spaces are much codified: women are only allowed to enter men's dining room to bring dishes and should leave their sandals at the doorstep.

Men start gathering in groups of three or four. Kiplangat and I are sitting with Ibrahim Oussa, a close friend of Abeid. And this is where the trouble begins. A woman gets into the room and brings us a first dish of *kisra* and *gurusa*[4] (delicious, but definitely stodgy food). As soon as we finish it, another woman brings the second one (see Figure 9.3). Same story for the third one. Sitting cross-legged on the carpet, I am now staring at my two dining partners. No problems with them: they are eating everything in sight, and more. I no longer feel hungry. I cannot eat anything anymore. It is very hot and stuffy in this room – heat of bodies,

Figure 9.3 Dinner scene with Ibrahim Oussa, Makina (Kibera).

Source: Jean-Baptiste Lanne, 2012.

tobacco smoke, smells of sauces. Should I implement some tactics? Giving some food to Kiplangat, hiding a bit of *gurusa* below my *kisra*? Ibrahim Oussa notices my little game. He tells me we should finish the food together. His voice is soft, but he is definitely not joking. It is a question of respect for the women who prepared the food, he says. It is about acknowledging the fact that we are doing something all together: celebrating Abeid and Zulaikha's wedding. Acknowledgement of a chain of production/consumption: sharing the same energy flow, or something like that. Enough with the analysis. I finish my dish.

Recording 3 – across Kibera, by night – around 3:00 a.m.

After the dinner, all guests start dancing in the front yard of the house. At the end of the party, the music turns from traditional to R'n'B rhythms, while dancers stop following *doluka's* dance steps (Nubian traditional dance) and start wriggling their hips to the sound of Rihanna's or Sean Paul's beats. Even some Luo songs are played, and hybridize the spontaneous choreography. Kiplangat finally suggests we go back home.

It is chilly at night. Kiplangat leads me through the labyrinth of Kibera's streets (see Figure 9.4). I am freezing: I get goose bumps on my arms. I feel tired and dirty.

Figure 9.4 A street in Kibera, by night (Makina).

Source: Jean-Baptiste Lanne, 2012.

At the corner of the streets: some burning piles of garbage (smells of meat, putrefied vegetables?). People set them on fire in the evening to save space. I suddenly remember something: I am currently in a dangerous place. I mean: since my arrival in Nairobi, I have always considered that Kibera by night was probably one of the most dangerous experiences. The energy of dance slowly goes down; the energy of fear rises. I feel it as an internal, physiological rhythm. Kiplangat notices my fear. He grabs my arm, and keeps it tightly in his hand. He guides me as a child across the muddy streets – 'right', 'left', 'ahead' – taking control of my defaulting body. 'Stop'. 'Careful with the gap'. He actually does not speak at all. He controls me by *contact*: putting pressure on my arm, pulling or pushing my shoulders. It is as if I had a spontaneous – quite mechanical – intelligence of the space I am navigating through. My body recognizes directions, where I should turn, what I should avoid.

We arrive at Kiplangat's home. Before we go to bed (shared with his two little brothers), he disappears for a while and comes back with something for me. It remains in my memory as one of the most delicate attentions someone has ever had for me. A set of two toothbrushes, just bought at the corner shop next to his house. What I understand of this present: a form of recognition, by Kiplangat, of the fact that this experience – spending the night in the slum for the first time – is not insignificant for me.

We brush our teeth outside, at the public fountain. We then go to bed, and I fall asleep right away. End of the recording.

Searching for a path: from the Cartesian subject to the machine

Too little or too much consciousness?[5]

Now what to do with these three little recordings? I would like to start by saying what they are not, or what they do not pretend to be. From this field material – this particular wedding, and some other events – I do not pretend to extract a general theory of what festivity is (in Kibera or anywhere else). Coming to Kibera, my initial aim was to contribute to the literature on festivity, which has been developed for more than 30 years from the early work by Mickhaïl Bakhtin on Carnival in western medieval and Renaissance folk cultures (Bakhtin, Pomorska, and Holquist, 1984) to contemporary theories that assume festivity as a consumer product (Gravari-Barbas, 2000; Greenhill, 2001; Getz, 2010; Blin, 2012). However, some days after being immersed in Kibera field, I had to face two main challenges concerning my own positionality. These two challenges deal with the porous and highly arbitrary boundaries I am spontaneously fixing between myself and the other (my surroundings: both material and non-material entities). The first of these two challenges is about the difficulties I directly encounter when trying to catch an object – festivity and festive practices – supposed to be outside me. Such a posture doesn't convince me: I do not believe anymore in the 'fiction of the objective subject' (Hancock, 2004, quoted in Blidon, 2012, my translation). What I witness, what I see in the field, what I feel, and what I provoke cannot work as unmovable markers to fix theoretical boundaries. I definitely cannot consider my field experience as an immobile painting, delimited by a rigid frame, offered to my perception. I'm living in this painting, breathing, interacting with different entities, displacing objects, and finally changing the whole configuration by my own presence. I cannot behave as if this confusion subject/object wouldn't exist.

Taking a scrupulous precaution towards the 'myth of objectivity' (Blidon, 2012), I now face a second challenge: what could be commonly understood as an excess of psychologism. Following Édouard Gardella (2003), I choose to call psychologism an experimental attempt by the subject to make himself intelligible to himself. This last challenge has been developed by a rich literature dealing with positionality issues. According to these authors, I have to know *a priori* who I am, how I will be perceived in the field, and what I will be able to perceive from the field (Rose, 1997). I have to ask myself about my own legitimacy among a community (Blidon, 2012) and what kind of mechanisms my presence will provoke (Anthias, 2001). I have to ask myself if I'm allowed to judge the acts and thoughts of my interlocutors (Scheper-Hughes, 1995). I have to be aware of my situated, highly political production of knowledge (Haraway, 1991). All of these issues, well developed by feminist/critical geographers, filled a theoretical gap in the 1990s and remain very stimulating to this day. Moreover, they are, from an ethical perspective, undeniably necessary. That being said, I agree with some authors who warn against an excessive questioning about the self (Bourdieu, 2003). Indeed, it sounds not illegitimate to ask if, by focusing on this range of questions, there is any risk of letting my 'object' run away. I am doing my fieldwork in Kibera, where a consequent number of *stimuli* are permanently exciting my attention. The social fact I am trying to witness (festivity) is a short-lived one. I try to adjust the scope of my eye, and it is gone. Someone invites me to enter a house reserved to women – it's now! – and I do not have enough time to ask myself: should I go inside? Am I legitimate to enter? What kind of consequences will my action have on other participants? Either I do things, or not. I do it mechanically because I desire it, because I get tired of hanging around in a crowded courtyard, because it 35°C outside, and I am definitely too warm. This spontaneous flow of desire is the only opportunity for me to catch what's happening here and now.

As an answer to these two challenges – too little or too much consciousness – I propose in the following words to decenter my reflection from my initial topic. Festivity should rather be used as a tool to explore the way I should marginalize my conscious subject in order to rethink my field experience. This marginalization process is a double one: it is about finding the relevant distance between the myth of objectivism (considering myself as an outside observer) and a powerless act of psychologism (putting blinkers on by continually focusing on my own positionality in the field).

The desiring machine as companion in my field experience

Too little or too much consciousness? I propose here to try to get out of this blind alley by experimenting the Deleuze and Guattarian concept of 'desiring machine' as it has been developed in the first chapter of *Anti-Oedipus* (Deleuze and Guattari, 1983). By referring to their work, I do not intend to replace a general theory (of any kind) with another. I would rather use this work as a companion in my own experience of the field. In other words, the question 'do desiring machines really exist or not?' does not matter at all. I make use of them – I mean, I consider them as true – for the time of my experience. They exist for a while as 'heuristic fictions' (Gardella, 2003). In the following reflection, I consequently understand the concept of 'desiring machine' as a specific posture which doesn't limit the subject to the consciousness, but rather considers it as precarious assemblage made of material entities (bones, flesh, skin), body fluids and flows, senses, physiological feelings (hunger), and reactions (goose bumps). All entities constituting this machine are able to interact on their own (not necessarily following an action of the consciousness) with other entities from the immediate surroundings (a material object, another body, the wind, the coldness, for example). In other words, the concept of 'desiring machine' is a way for marginalizing the role played by the consciousness in the range of productions and interactions operated by the subject.

By considering myself as a desiring machine among others (human and non-human ones), I choose to break with the Cartesian conception of the ego, which defines the subject on the axiom of a preexisting soul (Andoka, 2012). I start from the premise that nothing exists outside – and is at the basis of – my internal processes and external actions. Such assertions are no more allowed: *I do that because that's how I am. . . . I feel like that because that's how I am. . . .* Put another way, the question of the ego has no more ontological value (Mitchell, 2000). I am (does it make sense to write that anymore?) a machine: I act mechanically, answering to mechanical (physical, chemical) *stimuli*. I feel hungry/I no longer feel hungry, I am too warm/I am too cold, I feel safe/I am scared, something (somebody) attracts me/disgusts me. Material (oxygen, food, virus) and non-material flows (fear, empathy) come and go through my machine and connect it to others. 'It breathes, it heats, it eats. It shits and fucks'[6] (Deleuze and Guattari, 1983, p. 2). Such flows are called 'flows of desire'. According to Deleuze and Guattari, the concept of 'desire' should not be understood as a lack (Plato's canonical definition, see *The Symposium*, 200a–200e) but as a production process. (My) machine is connected to others by continual and unpredictable flows of production. During the wedding dinner, my hunger connects me to the group of men. My non-hunger disconnects me from the group of women: the flow of production (they produce – I consume) is broken. My desire to see Zulaikha in her father's house separates my machine from Kiplangat's one: each of them branches out into different directions. The fear I feel in Kibera's streets reconnects me to his machine: he grabs my arm and leads me mechanically to his house. The toothbrushes he gives me at home: materialization of his empathy flow. Food, voices, images, body contacts, saliva, objects, but also fear, hunger, sexual and non-sexual desires: the question is no longer where such flows come from but what kind of machines they assemble or disconnect. Thinking from a machine perspective means to focus on the whole chain of production. I stop asking myself: *Why do I fear? Am I legitimate to enter this house?* – I actually just fear, feel the goose bumps on my skin; I actually just enter the house, pushed by a non-identified desire. By doing so, the landscape I am able to witness extends to non-expected dimensions: all my desires, flows, sensations, emotions are working far beyond my consciousness to connect me to the different machines composing the field. What was formerly called 'I' is no more significant: '[P]roducing-machines, desiring-machines everywhere, schizophrenic machines, all of species life: the self and the non-self, outside and inside, no longer have any meaning whatsoever' (Deleuze and Guattari, 1983, p. 1)[7].

The geographer, as seismograph and poet

Staying machinic . . .

In the previous section, I tried to show that we should become machines, or better: that we should concentrate all our efforts to recognize us as machines. Such a posture invites us to focus more on the mechanism of the action in itself than on the entities connected by this action. Instead of asking about the boundaries between subject and object, self and the other, a machinic methodology tries to make the fluidity of the action unfold and express. I follow here the advocacy by authors from the Actor Network Theory for developing a science 'in the making' rather than a 'ready-made' science (Latour, 1987, p. 4; see also Law and Callon, 1988). Put in other words, the focus point is placed 'on the recording, not on the specific shape that is recorded' (Latour, 1990, p. 61). Being a machine is about letting the action (the record) be performed, avoiding interruptions, modifications, and short-circuits by my consciousness. Before going a step further, I would like to anticipate some ethical dilemmas such a machinic methodology might provoke. Being machinic – i.e. to marginalize my consciousness, to restrict it only as a part of my machine – is not about purely removing the consciousness from the subject. To clarify this point, I need to be precise on what marginalization means to me: it is about being able to keep a right distance between the actions that occur and an insatiable consciousness that always pretends to define itself as a cause or a receptacle of these actions. Marginalization is not about being unconscious and keeping the subject away from ethical issues which might appear in the field. By definition, a margin always can, at some point, recapture the core. There are of course some situations where the consciousness should come to the foreground, especially when actions in progress are involving ethical issues towards other human or non-human entities. In this chapter, I choose not to go further in this direction, but I resolutely keep this path open. Defining which ethical circumstances should lead to the recapture of the subject by the consciousness is a fertile issue that might be explored in the future works to come.

In a more practical way, I would now like to be more precise on what can be understood as being a machine in the field. It consists in a double action of stimulating and recording. Stimulating is about being aware of all possibilities given by different entities and circuits composing the machine. For example, the skin can capture coldness as well as well as a range of emotions (anger, fear). Hunger can be felt through the intestines, and leads to the action of sharing a meal. The act of breathing might give some information about the rhythm and intensity of the field. Eyes can witness an unexpected scene, and choose either to keep focus or to look away (in both cases, it will lead to reactions by other machines). All of these entities should not miss the opportunity to stimulate connections or disconnections with other machines, in order to expand the range of potential actions the machinic subject might provoke. However, in order to represent them (i.e., to avoid them disappearing with the flow of time), such actions need to be recorded. I call 'recording' this meticulous work of note taking that should not be taken for analytical work. Of course, this remains valid for pictures, sounds, or whatever else. I record on my notebook the intensity of my sensations and desires, atmospheres I perceive around me, temperatures I feel, but also facial expressions, speeds of the action flows. Nothing should be considered as uninteresting or shameful. This perpetual effort of non-intrusion of the consciousness – staying at the margin of the subject – leads to the introduction of the first of the two figures I would like to present here: the geographer as seismograph. A similar terminology has been recently used by Rachel Pain (2014) in the field of domestic violence. This metaphor, coming from the study of earthquakes, reminds us that some spectacular events are 'only the visible manifestation of long-term, much smaller movements of the earth's crust' (Pain, 2014, p. 129). Human consciousness get only what is highly visible and tangible: the earthquake. And what does acquire the knowledge about the whole chain of production? The seismograph, which started recording a long time ago. In the field, my goal is to become, little by little, a seismograph. I shall keep recording insignificant details, little continuities and unpredictable bifurcations. The time has not yet come to say what is relevant or not. So, I have to keep a distance from this pretended voice inside me: *Where is the earthquake? When is the earthquake?* Maybe something is trembling underneath, and my self-centred questions – my 'little fetishisms and paranoias' (Saldanha, 2005) – let it run away. I should avoid wasting such an opportunity and stay at the production level, where there is no place for a scrupulous consciousness. As Deleuze and Guattari put it: '[T]he subject spreads itself out along the entire circumference of the circle, the center of which has been abandoned by the ego'(Deleuze and Guattari, 1983, p. 21).

The time has not come to answer these questions: Where is festivity? What is festivity? For now, I focus on my apprehension to enter Zulaikha's father's house, the huge portion of *gurusa* in my plate, and the darkness of Kibera nights. I act in different ways, and it provokes new flows of production – sometimes frictions, sometimes revulsions, sometimes nothing at all. Is that relevant or not? Only the record matters. Who knows what will make sense?

. . . and becoming poet

I have tried to define my own path: recording, collecting, catching sounds, lights, voices, intensities. Neither attempt to classify nor to order anything. Always try, as far as possible, to stand at the margin of my analytical instinct – the threatening resurgence of the consciousness. However, such a posture cannot last forever. I cannot be satisfied with a mishmash of sensations, recorded in different places and at different times. What I had come to catch in the field was fluidity – in other words, a kind of intelligibility of a particular part of social life. Something that could answer this question: *How does it work?* And such an answer presupposes a dramatic thread to proceed.

It is precisely the time from which I consider myself free to reinvest my consciousness. Following Deleuze and Guattari's metaphor, I am now leaving the circumference of the circle, and I am reaching the centre. This last shift, from the affects to the consciousness, is permitted by a specific gap between two temporalities: recording time (machinic) and writing time (poetic). Such 'schizochronia' has been recently evoked by Myriam Houssay-Holzschuch (2014) to qualify the lived detachment between the time of fieldwork (1994 in the post-apartheid South Africa) and the time of family life, teaching, and writing. What Houssay-Holzschuch points out is this unavoidable *untimely process*: producing, diffusing a knowledge about something that does not exist anymore. This is her, my, our tragedy: She will never recover the atmosphere of Cape Town just after the first democratic elections, and I will never feel the heat during the wedding procession again, as well as the frightening labyrinth of Kibera nights. Only documents, notes, recordings remain. For now, we are at home, in our countries, in our districts, in libraries, classrooms, or whichever place, but elsewhere. Even the language is not the same as before. What is at stake is definitively different: We have to write articles, books, chapters like this one, in order to ensure our position in a specific academic field (Bourdieu, 2003). The whole range of emotions, frictions, epidermic sensations connected to the field is definitely gone. I have no choice but to assume this *untimely* character of my work. Now, what can be done with such a gap? What can I do, trying to catch and analyze something which simultaneously disappeared with the extinction of my machinic activity? I think – and this is the main argument of this chapter – that we should take advantage of this time interval, and dive deeply into it. We shall not act as if we were still in Kibera or Cape Town (myth of the objectivism), and neither shall we lament on ourselves and our incapacity to recover the past fieldwork (excess of egocentrism). In this in-betweenness, there is a path, or better: a desire. Such a desire is not a lack (lesson from Deleuze), but a wonderful opportunity for production. This is, precisely, a *poïesis*.

We, as geographers (*graphein*: to notch, or to write) shall not forget that we are – and remain – writers. We are embedded in an act of *poïesis*. We are facing a mishmash of places, times, sensations, feelings, voices coming from the fog of our memory as well as voices shouting from the kitchen next door (children calling for dinner?). Our role is to assemble these heterogeneous entities, to give them a common blast, to link them up in a kind of fluidity. We are definitely *untimely*; in other words, we have the power to recreate time. Maybe I do not remember exactly how warm I felt during the wedding procession. Maybe I cannot exactly recall my apprehension just before refusing, for the umpteenth time, the plate of *gurusa* a woman kindly offered me. Maybe Kiplangat never grabbed my arm during our little trip across Kibera by night? Did I invent it? And how to measure fear – the one I felt in the past time – now, at home, in front of my desk? Facing the betrayal of my memories, the figure of the poet – the second one that I would like to introduce in this chapter – can be of great assistance. I define the poet – or the geographer-poet – as an intentional posture of the subject, who remains conscious of the unavoidable failure he's carrying within himself. The poet believes in an intelligibility of the surrounding world, but remains convinced that such an intelligibility cannot directly merge as a simple translation from his senses (what he recorded) to his rational consciousness. Rather, the poet has to deal with a crooked path that leads to the *resubjectification* of the machinic recordings. This path includes memory lapses, deformations of the recordings, derangements of senses, and unpredictable coding errors. The common discourse would argue that such mistakes are an offence towards scientific values. The poet answers that these coding errors are not simply errors, but actually tell us a specific truth which could not be reached otherwise. The real illusion consists of believing that what has been recorded remains unchanged in our memories and notebooks as if it would have been stored in a bank vault. However, this material keeps working inside us, it changes, it becomes distorted, and this *in-becoming state* is a part of the reality, by the time we are trying to write about it. As poets, we have to be aware of this work of creation. The act of recentring the consciousness is about imagination, interpretation, distortion, exaggeration, or minimization of the recorded material. As Deleuze and Guattari put it, it is 'not imitation at all but a capture of code, surplus value of code, an increase in valence, a veritable becoming' (Deleuze and Guattari, 1987, p. 10). Filling gaps, inventing what is lacking, imagining what could have been, reconstructing the lost time embedded within our present situation: I argue that such a work should not be considered shameful; it is actually – for us, as writers – our necessity.

To conclude this section, I would like to clarify a bit more on what I understand as *geographer-poet* by giving some of its (non-exhaustive) assumed postures:

1. Assuming the 'I' and its internal distortions not as a lack of objectivity, but as an opportunity for creation;
2. Acknowledging that 'I' is a part of this world. Inside or outside does not mean anything. Consequently, staying at the margins of the egocentrism as well as of the object-centrism;
3. Keeping recording facts and discourses as well as facial expressions, sensations, intensities, rhythms, colours, temperatures, weather conditions. An entity must always be caught within its world;
4. Paying attention to the little events and minimal frictions, to the hardly visible or listenable: The earthquake can emerge from them;
5. In the field, making no hierarchical distinction between what has been (actuality), and what could have been (virtuality). Both shall be recorded;
6. Assembling 'real' persons from the field with heuristic figures, making their voices undistinguishable. Any attempt to let us believe that the Other could speak through our mouth is a lie;

7 Not concealing the material and psychological conditions of the writing situation: They are necessarily part of the production process and will have significant consequences on the final product;
8 Experimenting new forms of production (movies, poems, performances, creative writing, interviews with non-human entities). Specific forms prescribed by the academic field (Bourdieu, 2003) shall not be ignored but eroded at their margins.

Conclusion

As a conclusion to this chapter, I would like to summarize the machinic methodology I tried to develop through my particular field experience in Kibera. Being machinic is a double process which I have portrayed as two different figures together entwined in the subject: the seismograph and the poet. The seismograph is the figure which focuses its efforts on marginalizing the consciousness at the circumference of the subject. In other words, it's continually struggling against potential intrusions of its consciousness in order to let other machinic parts of the subject express themselves (physiological reactions, spontaneous emotions) and fluidly connect with other machines in the field. I am convinced that such a posture is a fertile way to expand possibilities of what can be recorded in the field. The second figure, – the poet –, comes afterwards and deals with the resubjectification of the machine. Considering that the accumulated material within the machine is a living one, deformed and modified by the action of time, we as geographers have no choice but to assume ourselves as creators. Following this poetic posture is about assuming an 'I', who has to cope with a material *in-becoming* inside itself. The geographer-poet recentres his consciousness as the core of his restitution work (writings, movies, performances, or whatever the shape it might take), but this consciousness doesn't work as a straight translation/imitation of the recording material in the present time. Rather, this process shall involve a certain sense of tale, a sense of poetic narrative where eventualities such as deformation, shifts in significations, memory lapses, and even exaggerations are taken for accepted steps of the protocol.

As a final word, I would say that such an advocacy for becoming a geographer-poet is not only an issue concerning our particular scientific and academic spheres, but is also about our connection with life around us. Becoming a poet is about reestablishing the spiritual fluidity between science and life. I argue – and it is probably not the only path – that a certain sense of tale, a sense of narrative, of poetry in our production, whichever forms it might take, gives better answers to the desire of people living around us. I imagine Kiplangat – the guy with the toothbrushes, remember? – perhaps a month, even a year after the wedding and our expedition by night across Kibera. We sit in a bar, drinking a shot of whisky, and he asks me what I have extracted from my research, what I have understood from the way they live and enjoy festivities. This is a crucial question, indeed: a question of reflection on himself and his community. What I would like to be able to provide to him is simply a tale, a well-constructed little tale, something fluid, something he could immediately understand, against which he could react, something he could criticize, mock, or simply not believe. It is his own right, after all.

Notes

1 Quoted in Spronk, R., 2011. *Sex, space, and sophistication*, SIEFF conference, Amsterdam.
2 Luos and Nubians are two communities living in Kenya. Luos come from the region of Kisumu (western part of the country, on the shores of Victoria Lake) and represent the second community in Kenya in terms of population. Nubians in Kibera are descendants from Sudanese soldiers, enrolled in the British Empire army at the beginning of the 20th century. They are Muslims and used to hold majority within Kibera.
3 I use the term 'recording' in a metaphorical way in order to insist on my machinic dimension when I was collecting field material in Kibera. Therefore please note that these 'recordings' are not coming from any electronic device or voice recorder, but are actually part of the field notes I took daily from January to May 2012.
4 *Kisra* is a very thin pancake and *gurusa* is a vegetable broth, sometimes served with goat meat. Typical of Nubian food.
5 In order to help the reader better understand my reflection, I would like to clarify a bit my terminology. I call *subject* or *machine subject* the whole machine constituted by material and non-material entities. Among the material entities, we can find bones, flesh, skin, organs. Among the non-material entities, there is the *consciousness*, but also a wide range of affects, passions, and instincts. As a consequence, the consciousness is only a part of the machine subject. This part can be marginalized (during fieldwork, for example), or recentred. This process of recentring the consciousness as the core of the machine, I call it *resubjectification* of the machine subject.
6 In French: 'Ça respire, ça chauffe, ça mange. Ça chie, ça baise' (Deleuze and Guattari 1983).
7 In French: 'Partout des machines productrices ou désirantes, des machines schizophrènes, toute la vie générique: moi et non-moi, extérieur et intérieur ne veulent plus rien dire' (Deleuze and Guattari 1983).

References

Andoka, F., 2012. Machine désirante et subjectivité dans l'Anti-Œdipe de Deleuze et Guattari. *Philosophique*, (15), pp. 85–94.

Anthias, F., 2001. New hybridities, old concepts: the limits of culture. *Ethnic and Racial Studies*, 24(4), pp. 619–641.

Bakhtin, M.M., Pomorska, K. and Holquist, M., 1984. *Rabelais and his world*. Translated by H. Iswolsky. Bloomington: Indiana University Press.

Blidon, M., 2012. Géographie de la sexualité ou sexualité du géographe ? Quelques leçons autour d'une injonction. *Annales de géographie*, 687–688(5), pp. 525–542.

Blin, É., 2012. Sport et événement festif. La ville à l'heure des marathons et des semi-marathons. *Annales de géographie*, 685(3), pp. 266–286.

Bourdieu, P., 2003. L'objectivation participante. *Actes de la recherche en sciences sociales*, 150(1), pp. 43–58.

Deleuze, G. and Guattari, F., 1983. *Anti-Oedipus: capitalism and schizophrenia*. Minneapolis: University of Minnesota Press.

Deleuze, G. and Guattari, F., 1987. *A thousand plateaus: capitalism and schizophrenia*. Minneapolis: University of Minnesota Press.

Gardella, É., 2003. Du jeu à la convention. Le self comme interprétation chez Goffman. *Tracés. Revue de Sciences humaines*, (4), pp. 21–42.

Getz, P., 2010. The nature and scope of festival studies. *International Journal of Event Management Research*, 5(1), pp. 1–47.

Gravari-Barbas, M., 2000. *La ville festive. Dossier de synthè`se en vue de l'habilitation à diriger les recherches.* Habilitation à diriger des recherches. University of Angers (France).

Greenhill, P., 2001. Festival, anti-festival, contre-festival, non-festival. *Ethnologies*, 23(1), pp. 13–21.

Hancock, C., 2004. L'idéologie du territoire en géographie: incursions féminines dans une discipline masculiniste. In: C. Bard, ed., *Le genre des territoires*, Presses Universitaires d'Angers. Angers, France.

Haraway, D.J., 1991. *Simians, cyborgs, and women: the reinvention of nature*. London: Free Association Books.

Houssay-Holzschuch, M., 2014. *Schizophrenia as method? Navigating Cape Town as a French geographer*. AAG Annual Meeting, Tampa, FL.

Lanne, J-B., 2012. *Fêter à Kibera : Pratiques festives et mécanismes de territorialisations*. Master thesis. Lyon: ENS Lyon.

Latour, B., 1987. *Science in action: how to follow Scientists and Engineers through society.* Cambridge: Harvard University Press.

Latour, B., 1990. On actor-network theory. A few clarifications plus more than a few complications. *Philosophia*, 25(3/4), pp. 47–64.

Law, J. and Callon, M., 1988. Engineering and sociology in a military aircraft project: a network analysis of technological change. *Social Problems*, 35(3), pp. 284–297.

Mitchell, D., 2000. *Cultural Geography: a cultural introduction*. Malden: Blackwell Publishers.

Pain, R., 2014. Seismologies of emotion: fear and activism during domestic violence. *Social & Cultural Geography*, 15(2), pp. 127–150.

Plato, 1993. *The symposium*. New York: State University of New York Press.

Rose, G., 1997. Situating knowledges: positionality, reflexivities and other tactics. *Progress in Human Geography*, 21(3), pp. 305–320.

Saldanha, A., 2005. Trance and visibility at dawn: racial dynamics in Goa's rave scene. *Social & Cultural Geography*, 6(5), pp. 707–721.

Scheper-Hughes, N., 1995. The primacy of the ethical: propositions for a militant anthropology. *Current Anthropology* [online] 36(3). Available at: <http://escholarship.org/uc/item/2xq430hc> [Accessed 4 Nov. 2014].

Spronk, R., 2011. *Sex, space and sophistication*, SIEFF conference, Amsterdam. Available at: <https://mecon.nomadit.co.uk/pub/conference_epaper_download.php5?PaperID=5836&MIMEType=application/pdf> [Accessed 6 Nov. 2014].

Repoliticization

10 Marginal attachment and *countercycling* in the age of recycling

Francisco Calafate-Faria

Discourses on the recycling of urban waste are usually articulated under 'unpolitical' eco-logics (Crenson, 1971). The pressure of looming environmental threats makes it harder to discern how recycling circuits benefit some agents and marginalize others. I argue that circuits of material recycling are assemblages of humans, technologies, streams of materials, and social relations that demand the discussion of notions of social justice and attention to uneven urban geographies. Contrary to what is communicated by the visual symbol of recycling, the circuits set in motion beyond each bin, commercial truck, package material, or council leaflet that displays it, are not simply flat, fluid, unilinear, and circular movements of waste from selective bins back to commercial shelves. If one follows the materials and the bodies that transport and transform them, one will encounter the transactions, attachments, interactions, hinges, mechanisms, and 'membranes' through which inequality is engendered and reproduced.

This chapter centres on the recycling circuits set in motion by informal waste pickers in Curitiba, Brazil. The city's strong recycling culture combines aspects of the urban South with models characteristic of wealthier cities in the global North. *Catadores* (the Brazilian informal waste pickers) are people who rummage through the city's waste in search of materials to sell. They form a group of highly marginalized people, from migrants recently arrived from rural areas to long-term inner-city slum dwellers, mostly with no formal relationship with the bureaucratic or welfare state. These informal recyclers are attracted and marginalized by the luring promises of urban recycling. In Curitiba, where, depending on who counts, 4,000 to 50,000 *catadores* operate, recycling is also closely connected to dominant discourses of efficiency and environmental sustainability. In 1989, this city developed one of the first campaigns promoting household recycling in the world. In the years that followed, partly because of its recycling model, Curitiba became known in Brazil as the 'first world capital city' and in international circuits of urban planning as a 'model of ecological efficiency' (cf. Rabinovitch, 1992; Hawken, Lovins, and Lovins, 2010; Schwartz, 2004; Irazábal, 2005; Moore, 2007; see also *A Convenient Truth*, 2006). What is not usually mentioned in the mainstream narratives about the city is that *catadores* are responsible for more than 90% of the recyclables collected. The international self-promotion of Curitiba as an environmental capital has made its formal recycling system world famous. At the same time, it has made its prevalent informal recycling under-acknowledged and under-examined. My ethnographic research on the recycling circuits of Curitiba during 2010 aimed at pulling back the curtain of promotional discourses. This chapter is primarily based on data resulting from this ethnography.

In order to understand the processes of marginalization and attachment that keep *catadores* both at the core of the recycling circuits and peripheral to the values they generate, I look at the rhizomatic configurations that hide beneath the *tree principle* portrayed in the official narratives about recycling. Taking inspiration in theoretical tools devised by Deleuze and Guattari (1987) and developed by Delanda (2006), Mcfarlane (2011), and Simone (2004, 2011), among others, I argue that the *deterretorialization* promised by recycling hides a strong process of *coding* (cf. Deleuze and Guattari, 1987 and Delanda, 2006), which marginalizes the very agents on which it depends. I use the Deleuzian conceptual toolkit to distinguish between promise, actualization, and potentiality in the following way. Modalities of urban recycling offer a moral connection with the city and the environment. In this way, they seem to open lines of flight from the detrimental effects of profligate consumption to a more sustainable urban economy. This is why several urban governments and activist groups have fought to implement efficient systems of recycling, involving different sectors of the population in this effort (Weinberg et al., 2000; Hawken, 2007).

Yet the recycling systems that emerged from these movements have foreclosed the benefits of this revaluation into a new coded territory, in which values are unfairly distributed across the city and across the globe. Here I take seriously the various social scientists who have considered the political dimensions of recycling systems by measuring them against values of social equality and global justice (e.g., Gandy, 1994; Weinberg et al., 2000; Pellow, 2002; Gille, 2007; MacBride, 2012). Thus I bring together these literatures grounded in the tradition of environmental justice with those focusing on material politics. For example, according to Minuchin, materials play an important part in 'articulating and mobilizing alternative urban imaginaries'. Materials can be 'a thread from which to reconstruct the staging of contrasting urban projects and sensibilities' (Minuchin, 2013, p. 258). These reflections result from the author's analysis of the importance of the use of concrete in Buenos Aires in the actualization of urban potentialities.

Just like new construction materials, the recycling of consumer waste offers opportunities for an urban politics that engages bodies, materials, imaginations, and collective endeavour into the construction of a fairer and more ecological city. A material politics that I call *countercycling* may, I argue, open new and more tangible lines of flight. Rather than being a sign of a disappearing past, informal waste pickers of the urban South may be in a crucial position to engage with transformative material politics. In order to investigate this possibility, it is essential to understand what happens along the routes taken by the various recycling materials.

This chapter is divided into four sections, each of them associated with one type of material. The first one – glass – will help us examine some common misunderstandings about recycling. The second – metal – will introduce the work of waste pickers and their articulation with the paradoxes of convergence and redistribution associated with informal infrastructures. In the third section, the paper circuits will show what happens in Curitiba in the transactions between *catadores* and the formal commercial circuits of industrial supply. The fourth material – plastic – will carry us out of the city from the dealers that buy from waste pickers to an industry that uses waste plastic as raw material. The chapter will end with the discussion of *countercycling* as a notion that can help us identify the possibility of new lines of flight in the revaluation of urban waste.

Glass tangles: objects shattered by lines of substances

The word *recycling* means different things to different people and in different contexts. In its most common usage, it refers to recovery of consumer waste for industrial production. In this sense materials are recovered after the object has been consumed, the form separated from the matter, the structure transformed in stream. It usually implies 'deforming objects before they can be reformed' (Gabrys, 2011, p. 138) so that what is made to circulate is materials or substances, rather than objects. Classic theories of waste revaluation look at discrete objects transported between stages of value and are, as such, inadequate for the understanding of this type of recycling (e.g., Thompson, 1979; Appadurai, 1997; Rathje and Murphy, 2001). Instead, I propose that we look at recycling as the circulation of materials into, along, through, and beyond objects and people.

The distinction between 'meshwork' and 'network' proposed by Tim Ingold reveals what is at stake here. The notion of network is associated with discrete objects being transported through linking lines between contained nodes. In a meshwork, the author argues, lines intersect in particular organisms or things (Ingold, 2011, p. 83). This insight is critical to understanding recycling circuits and, in particular, the work and lives of waste pickers. Recycling doesn't move used objects back to shelves where they can be used again. Rather, substances circulate from various origins, earth resources or recyclable materials, attaining temporary assemblage forms, at particular stages, as objects or things with the contribution of energy, technology, and social resources. As O'Brien points out, '[T]he means of permanently disposing of or removing waste creates its own opposite: the appearance of a temporary value in commodities' (O'Brien, 1999, p. 270). This understanding of exchange-value as fluid and temporary also allows us to look at use-value in a different way. A thing or object that is consumed, used, and discarded is a temporary assemblage of materials, technology, and practical knowledge that has diverse origins and can be disassembled and reassembled through subsequent processes. Circularity cannot describe this transformation.

Consider the example of glass. A bottle can be recycled through a circuit of exchange and transport that carries it back to be washed and refilled. This cycle can be repeated until the bottle breaks. I observed this process during my fieldwork in Brazil. In cafes, bars, restaurants, and retail shops, empty bottles were collected from consumers to be transported back to suppliers. In Europe this practice has long been a thing of the past. More often than not, glass bottles go into a recycling bin where they are shattered and enter a circuit that only partially results in new bottles. When it reaches the reprocessing factory, it is 'unrecognizable as glass . . . [a] mass of gooey, crunchy, whitish material' (MacBride, 2012, p. 23). This mix of different colours and grades of glass cullet with other contaminating materials is hard to sort and is most amenable for use in construction, rather than in the making of new bottles. According to MacBride, the contemporary industrial circuits of glass recycling are not economical or ecological. It is only the pressure to produce quantitative results of material recovery that fuels the investment in comingled collection and automated sorting systems, while at the same time making the recycling of glass more opaque to the public.

For *catadores* in Curitiba, who transport various types of recyclable materials in their individual vehicles (see Figure 10.1), glass is one to avoid. Glass objects break, cannot be compressed, and are heavy. Their relative value in the informal markets is very low, especially taking into account the difficulty in handling them. The proximity between materials and the human body also brings to the fore one important aspect of glass objects – when they break they threaten to cut the human skin. The only time I observed *catadores* dealing with glass bottles was at the Green Exchanges organized by the municipality (Figure 10.2). The Green Exchange is an iconic municipal programme in which a series of trucks come to the *favelas* every week with a scale and a few boxes of vegetables from

Figure 10.1 A *catador* with a full load of materials.
Source: Francisco Calafate-Faria, 2010.

Figure 10.2 Catadora taking some recyclables with low value in the recycling markets, like glass, to the Green Exchange.

Source: Francisco Calafate-Faria, 2010.

various surplus streams. The residents can bring their recyclables and barter for vegetables at a rate of 4 kg to 1 kg. Some *catadores* may use these moments to make something out of any bottles they have picked up. But glass is usually kept separate from the other recyclable materials collected in their daily work.

Catadores collect waste objects made of paper, plastics, and metal. Then they separate them in up to 50 different categories of materials. *Catadores*' collection and sorting is the initial input of energy into the process of retrieving materials from post-consumer waste. From this point onwards, materials travel through different circuits along which they are transformed into industrial raw materials. Catadores and professionals who work with recycling in Curitiba often refer to these routes as 'circuits of reverse logistics'. It is tempting to look at these circuits as networks connecting discrete nodes. In my view, such a reading risks missing what happens along the lines they travel, which is crucial to the whole process. It is this argument that I will continue to tease out in the following sections.

Metal convergence: aspiration and volatility

Brazil recycles 98% of all the beverage cans consumed in the country, more than any other country in the world.[1] This achievement results from the conjoined efforts of the residents of each of Brazil's 5,500 municipalities. An estimated one million regular *catadores*,[2] as well as occasional labourers, spring up at any public gathering of people consuming canned beverages, with their black bags, shopping trolleys, and a variety of other carriers ensuring that no aluminum can is wasted. Thus a reticulation of urban infrastructures constituted by people in action achieves better results than any planned system of provision anywhere in the world.

The efficiency and reach conveyed by the numbers obscures the convergence of the circuit. According to the Brazilian Association of Aluminum, 170,000 tons of the alloy, i.e., 70% of all the aluminum recycled in Brazil in 2010, converged at a reduced number of factories in one single city in southeast Brazil: Pindamonhangaba, 150 km from São Paulo. The recycling of aluminum cans epitomizes the ways in which *catadores* are articulated with the circuits to which they contribute. Their initiative and endeavour add up to a gigantic recovery infrastructure. Yet their individualized and dependent attachments render their positions extremely marginal to a circuit that concentrates gains.

Metals are highly recyclable materials, as there are industrial processes able to melt and reshape them with lower material losses. Yet metal recycling is far from a fluid and continuous circuit. Global markets affect the prices of metals in ways that reverberate powerfully across recycling circuits, impacting on waste pickers' lives all over the world. Commodities markets and speculative movements at institutions such as the London Metal Exchange, as well as other performative pricing mechanisms such as trade publications, make recycling circuits highly unstable. The lives of waste pickers become attached to uncontrollable forces generated far away, but whose consequences are extremely tangible.

One day I meet Tobias at Curitiba's 'Zero Point', the centre of the city's central square. He has recently resumed work as *carrinheiro*.[3] He had stopped for a few years after a drastic drop in the price of recyclables in 2008–2009. He says, 'The price of scrap metal plummeted. Cans used to pay more than 3 reais per kg; then *boom* [!] they came down to less than R$1!'[4] Tobias only works with cardboard and metal (mainly aluminum cans) and had a complex operation going when the financial crisis hit. He was working with seven pushcarts. He would chain six of them in strategic spots near the city centre and then move one around to collect material. He would work through the night loading up the stationary vehicles, with the help of other *catadores* he hired. When I find Tobias at the Zero Point, he is employing a *flanelinha*[5] to look after his pushcart while he moves around to collect material or goes to eat. He tells me how he had decided to move to Curitiba in 2007 after a visit to the city in which he heard about and witnessed first-hand the 'so-called recycling'. After the crisis, he went back to his small hometown in the countryside and was now returning to start all over again.

When *catadores* express bewilderment at price fluctuation, their accounts usually have a dramatic tone. Julio, for example, whom I meet and interview on a back street in a peripheral borough of Curitiba, tells me: 'Before people started talking about the world crisis, the prices were much higher than they are today. . . . Then there was a huge drop and only now are they finally recovering'. Since then, he says, 'I always fear when the price of some material is high, that it might be about to plunge again'. This feeling of powerlessness in the face of the fluctuations of the revenues of their work resonates with Bruegger and Knorr-Cetina's account of a different market: trading rooms in world financial centres (Knorr-Cetina and Bruegger, 2002). According to these ethnographers, traders perceive the market as a 'deep and liquid object', a 'greater being' with which they interact on a daily basis without ever managing to understand or control it (*ibid.*). Just as traders are often lured by the possibility of taming or manipulating markets, so are waste pickers often adamant of the possibility of short-circuiting or taking advantage of the circuits that exclude them.

In Tobias's story we see an example of how entrepreneurial strategies can be deployed to further one's position. Another way *catadores* seek to improve their situation is by assembling in cooperatives. Yet the process is not always easy.

Consider the case of Selma. She has been recently elected vice president of a newly formed cooperative. She will soon move to a new house next to the new coop. Meanwhile, because the new cooperative is not yet fully equipped, she stores the material in the area adjacent to her former house to be, in a *favela* by an inner-city river. In the area around her house, Selma and her husband show me the material they collect, just metres away from the middleman to whom she presently sells. Selma speaks in a quiet tone, but without whispering. She is clearly not worried that her neighbour may hear what she thinks of him:

> For this material here [cardboard], he [she points to the middleman's house] pays 30 cents [per kg]; there [through the cooperative] it will pay 45 cents. Sometimes I sell a ton at once here! You can see the loss I am making. Aluminum cans are worth R$2 [per kg]. There they will be worth R$2.5, or more.

We pass by their animals and the horse cart with which they collect. She's holding on to some materials until the cooperative opens. Then their materials will be assembled with the products of various *catadores*' collections, which will open access to dealers with more capacity than the middleman next door.

> I am keeping as much as I can, but I have to sell most of this material here because I need to live. But it hurts. This middleman next door used to be a bus conductor. He has been here for four years and already has five trucks, a house in the country, a house at the beach, and I, working for 16 years, do not even have money to buy a bike!

Aluminum cans are favourite objects for *catadores* as they are compactable, transportable, valuable, and abundant (Figure 10.3). As cooperatives are able to assemble and bale materials, they can transport higher quantities and get better prices. The shared use of means of transportation for materials also increases bargaining power by expanding the *catadores*' reach to more locations in the city and beyond. In some cases the buyers will still be informal, but they will probably be larger-capacity operators. This happens for aluminum cans as for other streams of material, and in the next section, we will travel with a load of paper from its collection and sorting by the members of a cooperative to its point of sale.

Figure 10.3 Aluminum cans sold by *catadores* being compressed at a scrap dealer in Curitiba.
Source: Francisco Calafate-Faria, 2010.

The paper journey: intersections and membranes

I am sitting in the right-hand side of a truck's cab (Figure 10.4). Through the mirror, I can see one corner of the load. Two neat piles of cardboard stand at the front. Behind it, coloured bits of different sorts of paper flap in the wind along with loose ends of plastic straps. We carry a relatively small load composed of cardboard, third-class paper, newspaper, coloured paper, and Tetra Pak. The total is 4.3 tons. It should pay R$1,490 to be distributed to the 40-something members of the cooperative.

Like the two men who sit next to me, all of the members of the cooperative live in the adjacent *favela*. As we drive out of the quiet unpaved street into an asphalt road, the two *catadores* recognize a man standing by a flashy sports car. 'That's X, from our *favela*. He's a drug dealer', they comment. The settlement is small but according to what they are saying, there are connections between the local dealers and a notorious armed organization run by prisoners in São Paulo.

As we leave the cooperative, the *favela*, and the drug dealer behind, we cover our first kilometres on the highway Green Line, a flagship work of the recent municipal administration. At one point in the 14 km trip, we pass by a large agglomeration of cargo containers. The area resembles a seaport, more than 100 km away from the sea. Here, containers that come from the Paranagua port to move into and through Curitiba are parked and inspected. We are crossing a crucial node in commercial and industrial routes. Passing large roundabouts and one flyover, we intersect a busy transport route with our load of waste materials assembled and sorted at the

Figure 10.4 The truck carrying paper from the cooperative to Delini.

Source: Francisco Calafate-Faria, 2010.

small warehouse next to the *favela*. Part of this material will end up composing new products that will travel in containers like the ones we see parked and piled in the large site by the road. Our truck looks much smaller when imagined as a point moving through these lines. It also becomes part of something bigger.

Recycling routes physically interweave circuits of transportation that feed industrial production, mimicking them, reproducing them, feeding back on them, and becoming part of them. Material for recycling is transported through the same roads as the raw materials that they replace and complement, the same routes as the objects from which they were extracted and the ones that they will constitute. These routes also carry the drugs and food that run through the bodies of people in the city. Thus, although some circuits are formally represented in the city's discourses and others are informal and made invisible, they share both material routes of transportation and biochemical cycles of human consumption. At the end of the road, I realize the importance of seeing the cooperative and the *favela* as crucial intersection of lines through which bodies, substances, and objects circulate and dwell. Then we arrive at Delini.

Delini is a large wholesaler of recyclables that deals with different sorts of paper, Tetra Pak, and pre-consumer plastics. It has an area of 10,000 square metres. It is a tidy and professionally organized space. The company has a fleet of ten trucks and employs 30 trained staff. Roger, my guide, is responsible for Delini's accountancy. As he describes all of the steps of the process, we walk through the spaces where paper is processed or stored. Delini buys from cooperatives of *catadores*, industries, the municipal recycling facility, and from informal middlemen. The latter buy their material from individual *catadores*, like Selma (see the previous section). Delini does not buy from individual *catadores*, as it needs regular and large intakes of material. As we walk, I comment that middlemen and wholesalers, like Delini itself, are the point of transition between the formal and informal markets. This seems to engage Roger, who responds:

> Yes, that's a good way of seeing it. . . . Because we have all the state licenses from the Environment Institute, everything is documented. . . . And when we deal with industries, we generally hand them a certificate of appropriate disposal, which the law requires.

Thus the *catadores*' cooperatives operate on the periphery of this regime of formal documentation. We stop in a space where the noise from the working of people and machines fades out into the silence of hundreds of piles of compressed cardboard and shelves of cardboard rolls. Changing to a slightly scolding tone, he adds:

> Aw, and speaking of which, taking advantage of your presence here, maybe you can help me, because we need associations to have all the CNPJ [collective fiscal number], to be completely regular. We need to start writing fiscal notes of material entry [tax receipts] 'against' them.

Roger explains how he struggles to include the transaction with cooperatives in the company's accountancy. There will be a register in the cooperative of this transaction, but there is no exchange of documents between the parties. Furthermore, Roger's appeal to me in order to bridge this gap in communication is also revealing. It signals that there is an important transition in the circuit happening in this transaction.

Coletto (2010) refers to the people who buy from *catadores* as 'membranes' between formal and informal sectors: 'As the recyclable material passes through this membrane, it acquires greater economic value' (2010, p. 80). Instead of providing mere sustenance, recyclable material starts generating profits. Furthermore, these 'membranes' act as filters, or permeable walls, between *catadores* and the official economy, allowing materials to pass through but revenues and particular kinds of knowledge to remain contained in specific parts of the circuit. These selective barriers help to explain how waste pickers are at once marginalized and attached to the circuits that they feed. The final section of this chapter will develop the concept of membrane and finally propose *countercycling* as conceptual material for new lines of flight.

Plastic politics: containment and *countercycling*

Tio Gerson (Uncle Gerson) was a name I heard mentioned in seven different cooperatives in Curitiba. He is a middleman who buys plastics from *catadores*. The quasi-familial relationship, which the nickname evokes, is an obvious part of the business strategy of this middleman. He buys more than 40 types of plastics from all 13 of the city's cooperatives, as well as from cooperatives in the rest of Paraná and neighbouring states.

The contrast between Tio Gerson's operation and the conditions at the paper dealer described above is striking. Gerson has three trucks, but one sits broken down at the back of his office. Another is full of baled material waiting to be unloaded. There is a conveyor belt for separation, which also lies inoperable and abandoned in the open air. 'Someone will come to fix it today', he explains. A good portion of the bales also lie in the open air, subject to dirt from the soil and deterioration from the weather. The impact of these conditions on the quality of materials will become evident in my later visit to one of Gerson's buyers. He invites me to sit in his office. He is drinking yerba mate from a makeshift gourd made out of a plastic bottle cooler. He explains,

> We've been in this depot for one year and a half. We used to operate in a farmhouse (*chácara*) kind of clandestine but for four years now we have been legalized. Well, my environmental license has expired last month – I have to do one or two things to get it back. But it is not a big deal.

During my first visit to a big cooperative in the centre of Curitiba, I heard its president agree a delivery with Gerson. After hanging up, he informed me that he went back to selling to him after experimenting with some buyers who paid higher prices. As he explained, 'They used to pay more but they are unreliable. Sometimes it is better to get a slightly lower price but to have security and consistency'. Some months later, in his office, Gerson confirms this point. This idea is central to his business ethos:

> I heard this week that a buyer tried to enter other associations, but they called me again, because they say that I offer them security. And also because I deal with them at a personal level. And of course there is the partnership. I have earned less and taught them how to separate. For example, now the prices are getting much lower and they will remain low until Carnival. There is a lot of material in the summer and the prices drop. Then it picks up. But I keep on buying, even if sometimes I lose money.

For this middleman, the reasons for his success amongst the cooperatives of *catadores* lie in his ability to provide security. He acts as a sort of welfare and insurance agent. This relationship configures a particular type of socio-economic relation, which in Brazil is often referred to as *paternalista* – 'paternalistic'. Other manifestations of this type of relationship between *catadores* and middlemen are prevalent when it comes to non-cooperative work. Stories of middlemen who regularly pay *catadores* in alcohol or crack cocaine, and/or rent accommodation to them in the premises where waste is stored are very common. Debt and other sorts of dependency between *catadores* and middlemen are mechanisms sustaining the marginal attachment of *catadores* to the recycling circuits.

In the previous sections we have seen other ways in which the links between *catadores* and the circuits of recycling tend to solidify attachments. *Catadores*' marginality and vulnerability is constituted by cumulative processes of interaction and attachment, rather than through occasional, dissipative, and exclusionary exchanges. It is the lure of the possibilities of rubbish, the paternalistic exchanges that engender dependency, and the articulation of convergence with dispersion that makes the integration of *catadores* with recycling selective and marginalizing. In the next subsection we will delve further into the circuit of waste plastic to get a sense of what happens to materials once they separate completely from those who first collected and sorted them. For that purpose, we follow a stream of plastic materials to a factory outside Curitiba.

Making bin bags out of recovered plastic: the other side of the membrane

Deep into the Paraná state, some 300 km away from its capital Curitiba, stands Mauri. In this small town of just less than 55,000 people, an eponymous industrial site makes bin bags and industrial wrapping film out of recovered plastics. Gerson suggested a series of clients for me to visit and I chose this one because it was the closest to Curitiba. Even so, in order to visit it and interview its owner, I travelled for more than five hours on a coach. Distance makes it even more difficult for *catadores* to circumvent the role of middlemen.

Sílvio is the owner and executive director of Mauri Plastics. In his office, 'entirely decorated with recycled materials', I find out that he actually stopped buying from Gerson more than six months ago: 'I am still opening the bales of material that he sold me; every time I do it, I have some nasty surprises'. He explains:

> Sometimes we receive bales of 200–250 kgs and when we open them there is 10–15% of other materials. And on top of that direct loss, there is the loss derived from contamination from organic matter, fat, and other plastics that are not mixable.

The problem, for Silvio, is that the middleman buys the materials and passes them on to the industry without doing any quality checks. Furthermore, the physical condition of Gerson's deposit means that materials are 'dragged directly on the soil and kept for days in the rain', which 'aggregates even more contaminant value to the materials'.

Silvio has visited some cooperatives, but it is difficult for him to do it more often because of the distance. Therefore, he says, 'the link, even of friendship, between the person who is there and buys from them every day is crucial'. The industries cannot buy from *catadores* directly until they manage to centralize distribution. Industries need to have large quantities supplied at regular times. The notion of 'membrane', to which I referred above, reemerges here. Not only because middlemen act as barriers to knowledge transference through the close personal relations they establish with *catadores* and endeavour to establish with the industry but also because their role as a conductor of material may assume a passive or even destructive character. The fact that Gerson suggested I visit Silvio's factory, even though he stopped buying from him more than six months before, suggests that he was trying to use me as tool of reengagement with his client. As a middleman, he uses personal forms of engagement to sew the circuit together. On the other hand, his position, as well as the geographical lag between the two sides of this circuit, erects a selective barrier to the circulation of information necessary to keep the flow of materials in good shape. Even though he claims that he endeavours to provide a constant stream of revenue to the *catadores*, in fact, his role in the circuit is accurately described by Colletto's notion of membrane.

Countercycling: lines of flight

As I have shown, attachments between people, materials, and circuits have created both a stable occupation and precarious experiences within it. I also want to argue that the unfulfilled promises of waste materials may still hold hope for these marginalized populations. The epistemic and practical capital generated by *catadores* may still allow them to open material routes of escape from cycles of exploitation as much as they offer aspirational energies. Entrepreneurial or even cooperative engagement allows *catadores* to attain slightly better positions in the market. They may also contribute to reproduce the forms of marginal attachment that characterize the occupation as a whole. Other examples offer more promise of superseding exploitative cycles. For example, in metropolitan Sao Paulo, one *catador* invented a machine made out of recovered materials that spins plastic bottles against a blade, allowing the extraction of a string of PET (polyethylene terephthalate) from plastic bottles into a yarn – also known as 'plarn'. The plarns (one from each bottle) are then interlaced to form a washing line. The interlacing machine is built with broomsticks and other objects found in the waste stream. The washing lines can then be sold to consumers, converting values of ecological production and social justice into economic value for the cooperative. This circuit is completely controlled by *catadores*, with most of the value it generates shared amongst them, rather than being drained into an industrial cycle to an industrial cycle. I found another example in a cooperative in metropolitan Curitiba. The members built a solar panel to heat water for domestic consumption by cutting and assembling plastic bottles into a structure that heats up water as it passes through the panel. Unlike most plastics, because it is transparent to UV rays, PET can be used to heat water.

The traits and potentialities of the material itself are important here. Designed as a raw material for textile manufacturing (polyester), the uses of this polymer resin have diversified. Like other plastics, PET was created and has been widely used for its plasticity, mouldability, and resistance to decay. In water bottles it has also been used for its lightness, portability, and, ultimately, disposability. Thus plastics' 'economic capacities are not so much intrinsic as enacted' (Hawkins, 2013, p. 49). This allows them to offer possibilities for a more emancipatory material politics (cf. Gabrys et al., 2013; Minuchin, 2013) but also illustrates that that they do not generate value on their own. Materials need to become part of particular assemblages deploying human will, ingenuity, and cooperation. Despite their precarity in the urban economy of recycling, *catadores* occupy a position that affords them the possibility of intervening through forms of material politics. This potential results precisely from the tensions between attachment and promise that at once reproduce both marginality and hope. Between the openings and foreclosures of systems of revaluation, between frustrating lines of flight and new recodings of the territory, those who circulate through the city and sort, collect, and engage creatively with materials defined by their potential move in spaces that may lead to unanticipated futures. Where recycling forecloses, countercycling may open.

Conclusion

Based on ethnographic fieldwork at the recycling circuits of Curitiba, Brazil, contextualized through analysis of the global construction of contemporary recycling, this chapter pulls apart different streams of materials to investigate how marginality is engendered. I argue that recycling, once a product of progressive campaigns aimed at overcoming a serious environmental problem, has reconstituted inequality and unsustainable productive models by recoding the territory through the unfair distribution of revenues and the marginalization of crucial contributors to the process. The crux of my argument here is that marginalization is reproduced through the attachment of populations to positions from which they make vital contributions to the circuit without correspondent participation in the distribution of the value created. Thus I am arguing against the view that marginality is reproduced through exclusion and detached occasional practices. An understanding of material assemblages is essential to this conclusion.

Methodologically, the chapter sought to engage with the material assemblages formed through the circulation of recyclable waste, in a move against the mystifying effects of the 'recycling utopia'. The recycling symbol and the correspondent unpolitical, abstract, and flat vision of resource efficiency, on the one hand, and the social science accounts that resist an engagement with the material as an intrinsic part of urban life, on the other, contribute to this mystification.

The previous section closed with an opening. The unfulfilled potentialities of waste materials include opportunities for those who occupy marginal positions in the urban economy. Countercycling movements of different sorts have potential to shape alternative futures. The subversion of systems of classification and the opening of possibilities for a more diffused power to define what or who is saved from wastage, and how and for whom, are the mechanisms that characterize countercycling. For those who provide materials to the recycling circuits, there is still a power they can exercise from their marginal – but pivotal – position. They are the ones who start the material assembly with their acts of gathering, sorting, and valuation. Thus they can open new lines of flight in an over-represented and over-signified model of virtuous circularity that depends on strong attachments at its margins.

But the conclusions of my investigation of waste pickers and recycling in Brazil are not limited to a sector of the population in one city or one part of the world.

In fact, waste pickers' position of attachment and marginality can also be used to theorize the position of consumers who participate in municipal recycling elsewhere. Aren't they also lured to contribute their free labour to a circuit of which they know very little (apart from mystifying representations and empty signifiers) and from which other agents extract the profits? And aren't these consumer households permanently crossed by streams of materials which they buy, sort, select, 'recycle', and redirect, but whose effect on their bodies and on their position in the urban economy and polity is always ridden with frustrating promises? Thus redirected to wider assemblages, the analytical power of these processes of attachment and marginality reveal the contrast between the virtuous circularity of the recycling symbol and the dynamics of power and inequality that unfold in the process of recycling waste.

Acknowledgements

I am deeply grateful to Rosemarie Paul, Margarita Aragon, Yari Lanci, and Michele Lancione for all the effort and talent they put into this chapter.

Notes

1 Source: *Compromisso Empresarial Para Reciclagem* (CEMPRE): http://www.cempre.org.br/.
2 *Ibid.*
3 *Catador* who collects with a pushcart (*carrinho*).
4 At the time of the interview R$1 = £0.33.
5 An informal street worker who helps park cars and cleans window screens (hence the word *flanelinha*, derived from flannel) in exchange for tips.

References

Appadurai, A., 1997. *The social life of things: commodities in cultural perspective*. Cambridge: Cambridge University Press.
Cetina, K.K. and Bruegger, U., 2002. Traders' engagement with markets: a postsocial relationship. *Theory Culture & Society*, 19(5–6), pp. 161–185.
Coletto, D., 2010. *The informal economy and employment in Brazil: Latin America, modernization, and social changes*. New York: Palgrave Macmillan.
Crenson, M.A., 1971. *The un-politics of air pollution: a study of non-decisionmaking in the cities*. London: Johns Hopkins Press.
DeLanda, M., 2006. *A new philosophy of society: assemblage theory and social complexity*. London: Continuum.
Deleuze, G. and Guattari, F., 1987. *A thousand plateaus*. London: Continuum.
Gabrys, J., 2011. *Digital rubbish: a natural history of electronics*. Ann Arbor: University of Michigan Press.
Gabrys, J., Hawkins, G. and Michael, M. eds., 2013. *Accumulation: the material politics of plastic*. New York: Routledge.
Gandy, M., 1994. *Recycling and the politics of urban waste*. London: Earthscan.

Gille, Z., 2007. *From the Cult of Waste to the Trash Heap of History: The Politics of Waste in Socialist and Postsocialist Hungary*. Indiana University Press.

Hawken, P., 2007. *Blessed Unrest: How the Largest Social Movement in History Is Restoring Grace, Justice, and Beauty to the World*. Penguin.

Hawken, P., Lovins, A.B. and Lovins, L.H., 2010. *Natural capitalism: the next industrial revolution*. London: Earthscan.

Hawkins, G., 2013. Made to be wasted: PET and topologies of disposability. In J. Gabrys, G. Hawkins and M. Michael, eds., *Accumulation: the material politics of plastic*. London: Routledge, pp. 49–67.

Ingold, T., 2011. *Being alive: essays on movement, knowledge and description*. London: Routledge.

Irazábal, C., 2005. *City making and urban governance in the Americas: Curitiba and Portland*. Aldershot: Ashgate Publishing, Ltd.

MacBride, S., 2011. *Recycling Reconsidered: The Present Failure and Future Promise of Environmental Action in the United States*. MIT Press.

Mcfarlane, C., 2011. *Learning the city: knowledge and translocal assemblage*. Chichester: John Wiley & Sons.

Minuchin, L., 2013. Material politics: concrete imaginations and the architectural definition of urban life in Le Corbusier's master plan For Buenos Aires. *International Journal of Urban and Regional Research*, 37(1), pp. 238–258.

Moore, S.A., 2007. *Alternative routes to the sustainable city: Austin, Curitiba, and Frankfurt*. Lanham: Lexington Books.

O'Brien, M., 1999. Rubbish power: towards a sociology of the rubbish society. In J. Hearn and S. Roseneil, eds, *Consuming cultures: power and resistance*. London: Palgrave Macmillan, pp. 262–277.

Pellow, D.N., 2002. *Garbage wars: the struggle for environmental justice in Chicago*. Cambridge, MA: MIT Press.

Rabinovitch, J., 1992. Curitiba: towards sustainable urban development. *Environment and Urbanization*, 4(2), pp. 62–73.

Rathje, W.L. and Murphy, C., 2001. *Rubbish!: The archaeology of garbage*. Tucson: University of Arizona Press.

Schwartz, H.H., 2004. *Urban renewal municipal revitalization: The case of Curitiba, Brazil*. Alexandria, VA: H. Schwartz.

Simone, A.M., 2004. People as infrastructure: intersecting fragments in Johannesburg. *Public Culture*, 16(3), pp. 407–429.

Thompson, M., 1979. *Rubbish theory: the creation and destruction of value*. Oxford: Oxford University Press.

Weinberg, A.S., Pellow, D.N. and Schnaiberg, A., 2000. *Urban recycling and the search for sustainable community development*. Princeton: Princeton University Press.

Other sources:

A Convenient Truth: Urban Solutions From Curitiba, Brazil, 2006. [Film] Directed by Giovanni Del Bello.

11 The 'differentiated countryside'

Survival strategies of rural entrepreneurs

Eszter Krasznai Kovács

Introduction

This chapter outlines how three different groups of people – young men, gypsy workers, and small-holder farmers – try to 'make ends meet' in the rural Hungarian countryside. Rurality connotes a seemingly self-evident physical or isolated removal from the hustle and bustle of opportunity possible in better-networked towns and cities. Despite strong rural-to-urban migration and high outright emigration rates in Hungary today, more than half the population continues to live – many *choose* to live – in 'rural' areas. Herein I recount the labour and economic decision-making of such men and families from the Northern Cserehát, which along statistical measures is considered marginal and depressed: the Slovak-Hungarian border region is the economically poorest of Hungary, far from urban amenities and reliable transport networks and markets. These men attempt to take advantage of the relative neglect of the state to try to get by and more – to get ahead, to be comfortable, to claim and continue to make the 'rural' a viable livelihood option (Figure 11.1).

These struggles play out through outright dependencies on state support and presence – in the forms of welfare, compulsory work programmes, regulatory oversight, and agricultural subsidy support schemes – and state absence that enables a localized and niche 'leniency' that gives rise to a rich plethora of strategies that binary legal/illegal, licit/illicit, formal/informal categorizations fail to explain (Centeno and Portes, 2006). Attention to the complexities of the myriad personal 'moments' in time that individuals interact with the state highlights the differential risks, opportunities, and effects of state laws and powers on people, which frequently have unintended local consequences and biases as well as mask how the state is a willing participant, enabler, and producer of deviance (Saitta, 2013).

Government policies of both the EU, as well as contingently and independently the Hungarian state, encourage the 'fostering of dynamic entrepreneurship' of the countryside's inhabitants to incentivize individuals to create their own opportunities for employment and business, to take advantage of 'reforms . . . which have created a market-oriented environment' (EC, 2005; North and Smallbone, 2006). In the rural setting, this predominantly takes the form of tourism, farm renovations, and the seeding of financial support to small businesses for local produce, food,

Figure 11.1 A 'European Village' signposted outside Bodvalenke in the Cserehát.
Source: Eszter Krasznai Kovács, 2014.

or handcraft. 'Projectification' and the strengthening of individuals' capacities for enterprise are largely in keeping with a 'neoliberal' ideology and political agenda that encourages individuals to be responsibilized for all aspects of their lives alongside the withdrawal of state support (Castree, 2010). This pursuit of individualized success contradicts many aspects of formerly socialist societies, which were strongly centralized, with no official unemployment and agricultural collectivization (Hann and Sárkány, 2003). Thus the unequal ability of groups and individuals to participate and perform in this entrepreneurial game is inflected with its particular *rural* Hungarian historical, social, and political context, within which the state is broadly divesting and withdrawing institutional and infrastructural support.

The 'differentiated countryside' (Murdoch et al., 2005) recognizes how varied social and economic drivers of change impact different types of rural areas (Lowe and Ward, 2009), where global concerns are grounded and contested in particular locales (Woods, 2012). Recognizing such heterogeneity of place in turn complicates the expression, implementability, and even perceived legitimacy of law, regulation, and the over-arching 'place' of the state (Braverman et al., 2014). Here I turn my attention to such a site of contestation, where the socio-economic and physical (isolated) marginality of the Northern Cserehát has given rise to a suite of economic approaches and survival strategies in response to limited opportunities for making a legitimate, state-sanctioned income. These are undertaken by men[1] who frequently reject outright as misplaced and inappropriate the applicability of laws that were passed elsewhere, to regulate and tax businesses and individuals of distant, removed urban environments (Pruitt, 2014). Amongst respondents, expressions that questioned the authority of law were commonplace, such as 'they have no idea, sitting behind a table in Budapest'; the wishes and wills of the EU were interrogated as the EU was seen as composed of sites of wealthier, more 'well-to-do' Western countrysides such as Germany or the UK that had historically and repeatedly failed to represent or improve conditions in Hungary.

In the sections that follow, I make an attempt to speak of the survival and livelihood strategies along various individual, community, and group categorizations. First, I recount through personal narrative my acquaintance with two young men who work together on a farm but seek to augment their income sources and cash flows through additional informal work. Second, I describe gypsy[2] work brigades made up of Roma men who voluntarily form bands for manual work that can be completed together; these brigades predominantly comprise individuals on welfare who need extra income (cf. Boels, 2013), and who gain opportunity to make the most profit from the complicity of state forestry officials. Third, I engage directly with a sector propped up with state subsidies: small-holder farmers on agriculturally marginal lands. The role of the state and its (in)ability to 'capture' the incomes, activities, or aspirations of its rural inhabitants is central to all three accounts. However, these narratives also portray the very real stumbling blocks that a particular set of skilled and determined people encounter in their efforts to survive comfortably and in their own configurations and aspirations, to 'move up' in the world. The barriers to such movement are not all externally imposed, but relate to the normalized practices of the communities in which they are embedded, their social expectations, as well as a result of maintained nepotistic logics and practices that sustain local economies.

I moved to the Northern Cserehát in 2011, to undertake fieldwork amongst farmers and conservationists for my doctoral dissertation; over the years, I kept returning to the Cserehát to be with my partner, who is a farmer. Through this time, my research and positionality have evolved such that I was increasingly exposed to the decisions, strategizing, contradictions, and maneuvering required for livelihoods and survival. I have participated keenly in the running of farms as well as in community events and activities, keeping a field diary and conducting open-ended exploratory interviews to understand the relations and people around me. Below, I have made an effort to recount several of the perspectives that I encountered in narrative form, where inevitably the voice is my own, but fused with the descriptions and motivations as recounted to me over time by particular individuals, whose names have been changed.

Youth apart

A rablóbol lesz a legjobb pandur – 'Robbers make the best gendarmes'.

Zsolt and Miki are cousins, 28- year-old men with wives and children at home. Their fathers are long dead, in keeping with the low male life expectancy of the region, and they have respectively inherited small trouser strips of land that they sow and reap by hand and horse. Two years ago Zsolt decided to lease out his inheritance to the largest local farmer because 'everyone' around him was. From the leasee he now gets a percentage of the agricultural subsidy in the form of wheat or oats, not cash, symptomatic of their perpetual general struggle to receive 'proper' payments. Struggling to maintain cash flow binds the landed and unlanded, rich and poor. Zsolt and Miki's childhoods and identities are self-proclaimed as rural: They claim to have no urban aspirations or desire to seek work in the city or abroad. The country is 'good enough' for them and it's all they've ever known. They have never been farther away than Budapest, four hours away by car. They have never voted and express active disaffection from national and local politics as 'things are what they are, and giving a damn would not change anything'.[3] They are both enterprising and very skilled with their hands. Miki prefers to work in the forest, amidst the sharp cold and tang of pine, to be outdoors all day. Despite the hard physical work that goes with cutting down trees, he is already growing a compact paunch and has ever-larger and ruddier cheeks in the three years I have known him.

On paper both men are employed by Attila, who is a local land and forest owner. This means that they get insurance, social security, and pension payments made to the government in their name. They are declared at minimum wage, around 66,400 Hungarian Forints (HUF) net per month (US$249), but in reality take home a daily rate for farm work, or by the cubic metre for cut timber if working in the forest, resulting in a real cash-in-hand income of around 120,000 HUF (US$450). This way, Attila doesn't have to pay the extra pension and security payments. Originally they were employed five days a week; however, Attila feels that his business is increasingly precarious as a result of political and legal uncertainty: the threat of more taxes, deflated and fluctuating markets for goods, alongside an ongoing weakening of the currency. Thus, he has cut his men's official employment hours. When he registered them to work at officially only four hours a day, Attila had to pay 26,300 HUF (US$98) in compulsory payments after them to the state, even though Zsolt and Miki only took 33,400 HUF (US$125) cash-in-hand home – Attila had to pay out almost the men's wages again to the state to comply with their required social securities. Attila feels that these extras to 'the state' (he does not see them as his employees' long-term superannuation, but as an attempt of the state to exhort more money from his business) were insupportable after half a year, so he further reduced their official hours to two hours a day (after which he only has to pay 12,750 HUF, or US$47, into pension schemes). In this way, the men are engaged in a semi-formal arrangement that benefits our protagonists' short-term interests with more cash-in-hand as well as savings for Attila. Zsolt and Miki never thought to argue about their reduced pensions, as they feel lucky that they continue to have any declared income at all.

Miki and Zsolt naturally never really worked only two or four hours a day. Both men are invaluable on a farm, as they can do everything. When and as required they work in the stables, deliver calves, cut wood, sit in the tractor to plough, mow, or make hay bales. They are keen and frequently work seven days a week, but they have been unable to save any money over the past three years.

Miki's occupation 'on the side', after extracting the wood for Attila, is to transport and sell it on with his 'cut'. Attila sells him the wood and Miki transports it to neighbouring villages, charging for the transport, for the packing and unpacking of the vehicle, adding 10% to the cost of the wood. In the villages, people get paid on the fifth of the month, whether working privately or in welfare, so he is most active as we speak now and as I write, on 7 February, working against the cold.

The widespread lack of cash or liquidity prevents the poorest households from even short-term planning. Through Miki, people buy pine, which is the cheapest wood. However, it burns the fastest and has the lowest heat capacity. His customers thus work their ways through this wood the fastest, a lack of cash preventing them from buying a mixture of oak-poplar-pine, a blend of hard and soft wood that burns longer and gives more heat but costs double. The journey of the timber turns illicit once it gets loaded onto Miki's trailer, as Attila provides Miki with a transport receipt to verify that the wood has been cut with permission; from then on, the wood's journey is not further taxed or captured.

Symptomatic of an aversion to state regulation and bureaucracy, both men's papers are not 'in order'. Limiting Miki's enterprise is his lack of a driving license. He has driven since he was a teenager but never completed a course due to not seeing why he needed one – different priorities, a constant shortage of money, and now a level of pride that he has lasted so long without one. *Mi az nekem, én tudom hogy mit tudok* – 'what's it to me, I know what I know', he replied when I asked him why. Now that he has upwards of ten priors for being caught by the police for driving without a license, he cannot start a course easily. The local police have learned to spot his vehicle and stop it, expecting to find him behind the wheel. Thus, Miki's business strategy has to include the hiring of a driver, usually a mate, who has a license and can be relied on to drive the dilapidated car, which doesn't have a handbrake, and be able to cope with an engine that cannot be stopped, as restarting it requires an external electrical surge. His efforts at enterprise demonstrate well the hodge-podge nature of his instruments, his preparedness, the (seasonal, infrequent, unpredictable) demand, as well as the extra financial costs he must incur as a result of his administrative illegitimacy.

Zsolt works with wood once it's been chopped down, in sawing and carpentry. His home is surrounded by the artefacts he has made – a wooden bridge, bird feeders, and postbox. On the side, he takes odd-job orders that drip in sporadically from surrounding villages, to make wooden seats for summer beer festivals, for example. Most of his clients do not pay him ever, and those who do pay him late. Zsolt is ever the optimist, however, and thinks new orders will result in ready cash-in-hand, even with individuals who have owed him for other jobs for years. Zsolt and Miki both work, decidedly, without invoices, receipts, or any paper trail, and the people who hire their labour or buy Zsolt's efforts do not ask for them. This *láthatatlan gazdaság* or 'invisible economy' is entirely normalized amongst individual labour-for-hire in the Cserehát. Zsolt has been unable to take on a larger order from a school because of their requirement for him to provide them with an invoice; his refusal arose from an unwillingness to start the process; *mit hiányzik ez nekem* – 'this is not missing from my life': to be registered, seen, and captured by the state's tax authorities. Such an order has only come his way once. Zsolt expresses a mixture of passive and active resistance to regulation and 'being seen'. The transaction costs of becoming set up and what this represents beyond having to pay tax – the requirement to let the state in – is seen as too high a price in exchange for state-sanctioned legitimacy.

Cigány brigades

Béla, Bélu, Gabi, and Laci have formed what is known locally as a *cigány brigád* – a 'gypsy brigade' – a work group willing to take on any manual labour in exchange for undeclared cash to supplement their welfare payments. In the mornings, all four men report to the local council office, as they are *közmunkás*, 'public workers', which is the Hungarian government's compulsory work for the dole scheme. This means they have to show up for work every morning at 7:00 a.m. during the winter, and at 6:00 a.m. during the summer, to be put to work with activities such as litter gathering, de-clogging of drains and gutters, planting of vegetables on council land, cutting grass on council territory, planting flowers, painting walls, and fulfilling any other general odd jobs that the local council can think up for them to do. Such work is done entirely by hand. There are many more people employed in *közmunka* than there is work available, for a start, so there is little point speeding up their productivity. For example, during the week seven people may be loaded into the council-owned Ford microbus to saw branches off trees along the sides of the road with a crocodile saw; the workers light a fire for warmth and for stew in the early morning and progress around 50 metres in the course of a day, idly sawing. In the late spring, dozens of workers are sent into potato fields to sow; here, groupings of *közmunkás* alternate between toiling and chatting in the shade until they are permitted to return home.

In exchange for this time, people receive the state welfare benefits that they had, pre-2011, obtained without working for the state.[4] Village populations enrolled in the *közmunka* programme count towards employment statistics at the national level, effectively lowering Hungary's official unemployment rates (MTI, 2011; Hír24, 2014). Alongside minimum wage checks, child and disability support allowances are also affected and potentially penalized if the *közmunka* programme is not attended. Amongst village communities *közmunka* is held as the route to 'sure money' in a rural context with little to no alternative work opportunities.

As Béla expressed to me, 'It's the only way most people can get any money. If they ask us to jump, we'll jump'. Communities are small, and failing to show up to *közmunka* has resulted in councils threatening to withhold free fuelwood and onion or potato allowances, even though they are legally not permitted to do so.

Thus, with the mornings taken in menial unskilled work for the state, these self-formed brigades are free to work in the afternoon. There is almost one *cigány brigád* in every village of the Cserehát, if not more. They are typically composed of one leader plus three to four men. With no capital, they cannot afford tools or cars, or petrol for chainsaws, so all the majority has to sell is their labour. Their work must be illicit and uncaptured by the state, as otherwise their *közmunka* payments would be threatened; but on the flip side, because of their lack of capital to invest in tools and the volatile and unpredictable rates of finding lasting work, there is no incentive to leave the *közmunka*. Most brigades find afternoon and weekend work in the forests through the state forest department, which ironically turns a blind eye to their 'illegality' for a fee. The extraction of undocumented and illicit wood through the collusion of state foresters and brigades depresses the local price of timber, in turn affecting the private landholders in the region.

In a forest brigade, typically two people cut the tree, one the branches and trunk while the other chainsaws these into logs that the further two members stack and pack on to a horse-drawn cart or dilapidated trailer. Brigade leaders have to be crafty to make a profit, easiest done when working within state forests. In these situations, they get told by state foresters to cut 100 cubic metres, as an example. When it comes to getting paid, brigade leaders have a term, to 'solve it smartly' (*okosba megoldjuk*), which refers to all types of solutions that are outside the law – including providing no paper trail, selling items without tax added, providing goods of variable quality, and so on. In the forests, the 'smart' solution is typically to cut 120 cubic metres for the official 100 cubic metre order, of which the 'spare' 20 cubic metres is sold and distributed without any official trace, from which the brigade takes its pay and the (corrupt) governmental forestry officer takes his 'cut'. However, brigade leaders have been known to attempt to spill over these practices to the private sector, offering the *owners* of private forests *themselves* a 'cut' of their illegally chopped-down plus 20 cubic metres' equivalent of wood, which is met with incredulousness by private foresters.

Private landholders are constrained from hiring brigades primarily due to issues that may arise as a result of brigades' illicitness, and their unwillingness to invest in providing groups with their own equipment and tools. Many brigade leaders resort to asking friends or local shop retailers to provide them with an invoice to enable them to take on work for a farmer, but this frequently comes with giving a 'cut' to the invoice giver, inevitably a sizable proportion of any profit. Furthermore, brigades must be paid very frequently to keep them in liquidity, which many farmers do not appreciate. Cash is required to maintain tools and equipment, to put petrol in the chainsaws, and the like; decade-old equipment in turn begets here a vibrant knowledge economy around repair. Russian-era machines are highly valued because of the ease with which they can be fixed, borne out of decades-long familiarity and endurance, and because the intricacies of their parts can be readily accessed.

Many individual *cigány* men, while not their brigades, are held by local farmers as worthy of praise as a result of their 'hard work' and stamina or individual 'entrepreneurial spirit' (*vállalkozú szellemű*). Amongst the farming class, there is respect granted to those able to drive a hard bargain. If forest work is unavailable, brigades turn to farm work, although this is too seasonal to be a reliable source of income. An ironic observation I made very early in my PhD fieldwork in 2011 was that the majority of people in the Cserehát are indeed *cigány*, yet there are no *cigány* farmers. However, brigades and individuals have been employed in significant numbers to work the land, physically charged with clearing large areas of sloe brush to convert areas into ploughable fields; once converted, through vast quantities of sweat and feats of endurance under the blistering sun, these same fields will yield them no more work. The large farm-holdings capable of hiring them in the first place are entirely mechanized when it comes to ploughing and tilling the land once it is cleared of root systems and stones. With widespread mechanization, only work at harvest time remains. Even the making of square hay bales, which employs two people beside the tractor, is on the decline; the advent of cylindrical bales requires only a tractor driver. In this way, there are diminishing opportunities for work even within informal 'shady' practices and connections.

The manifestation and relationships to the fragmented scales and forms of 'the state' must be remarked upon within these tales. Individual rural dwellers in the Cserehát, whether as landholders or as part of brigades, are dependent on the state for a base income, which both keeps families afloat at a barely tolerable level[5] while also constraining their ability to seek other work. In the *közmunka* programme, however, it seems likely that without seeking out illicit activity, these individuals would otherwise remain forever 'at the bottom' on minimal wage, unskilled. *Cigány* brigades are not a new phenomenon, as groups of men have banded and laboured together for decades and are an enduring assemblage of individuals in the landscape. These men actively elect to form in response to minimalist conditions, while their degrees of licit or illicitness varies with the political winds. Within the current state architecture of work-for-the-dole, social and economic development are not forthcoming, while additionally, most *cigány* brigades are unable to meaningfully capitalize on their 'extra' work. This is due to another layer of the state encountered in their 'everyday', in the form of policemen and local council officers. Policemen have been accused of instigating an increasing number of race-motivated sanctions. Béla, our friend from the neighbouring village and the head of the local brigade, has been fined for walking in the middle of the road instead of on the pavement for a cool 10,000 HUF (US$35), and also for driving his vehicles on a private dirt track, for 30,000 HUF (US$106). Most of his brigade members have expired licenses or car registration papers, a source of further fines, as they are invariably caught. These fines take away the savings an individual member of a brigade can make with his additional work, leading to a lived 'vicious circle' of poverty. I contend that these perpetual contradictory 'lose-lose' relationships with the state, combined with the pettiness and inconsistent application of the smallest rules with large fines, cause many rural individuals to not only feel an antipathy and mistrust to the state, but that this in turn results in the hollowing out of their expectations and ambitions.

Small landholders

There are many small-holder families in the Cserehát who have 'always' lived in the area. The majority assess their farming businesses and practices as financially untenable without the help of subsidies (Kovács, 2014). However, many do not farm out of a motivation to make a profit, but because it is something that their families have always done; many state that they will continue to farm until they run out of family members. The possession of land, and thus some capital, empowers these families to define a relation with the state that is of their own choosing. Many exercise this by electing to *not* continue participation in agri-environmental schemes or agricultural subsidy programmes that were designed to 'support' the very maintenance of farms like theirs and contribute to 'rural development'.

In other work, I have documented how subsidies have been inflexibly and bureaucratically administered such that small-holders have felt disadvantaged in their ease of participation for the subsidies relative to larger landholders (Kovács, 2014, 2015). As a result some farmers have come to question and resent the regulatory ambitions and oversight of the state that 'maintains' them. A representative view that I encountered at Hidvegardó, the last village before the Slovakian border, was from a 40-year-old farmer who had elected to withdraw his participation from high-nature farming after five years:

> We just want to be left alone! Farming here has always been hard or close to impossible, we could never live off the land, not at this size. So when this scheme came in [subsidies], we hoped that it would help. Instead, it brought inspectors, we had to hire accountants and farm administrators, we were fined three years out of five for our paperwork not being completed correctly or submitted on time. . . . After all of it the farm is still not profitable, I actually think we're worse off. [. . .] This is not how you do conservation! What has this got to do with conservation?

The 'traditional' farming form in the Northern Cserehát is on thin strips of land, called *nadrág szíj* or 'trouser strips' of plots. These small plots give rise to a mosaic landscape, with different crop types and timings reflecting heterogenous ownership patterns, which have been found to correlate with higher levels of biodiversity (Kleijn et al., 2006). The nearby Aggtelek National Park Directorate has demarcated and established an 'environmentally sensitive area' for the Northern Cserehát, as a result of several species' presence, such as the threatened corncrake and fritillaries. These incentivize 'environmentally friendly' or green farming within their boundaries through restricting fertilizer and pesticide applications, crop rotations, and even access to land during bird nesting season. However, the persistence of 'trouser parcel' agriculture in the Cserehát may also be an indicator of socio-economic poverty as well as agricultural marginality, with the two necessarily in a tightly bound relationship. Below, I recount three attitudes and points of relation to the agricultural support system.

A significant portion of 'traditional' farmers are ageing, approaching 70 or more years of age, and have farmed on these parcels their entire lives, and have typically also kept cows, pigs, and chickens. Jani bácsi, for example, claimed subsidies for less than two hectares of land that lay behind the house where he and his wife had lived for more than 50 years. Their two children had migrated to Miskolc and Budapest, with no intention of returning to the country. Jani bácsi had difficulty understanding the purpose of the subsidy scheme, as he had never used pesticides or aimed to mechanize his farm in any way – this would only happen if he were no longer able to continue, in which case his children would sell the parcel to a larger farm-holder, who would swallow his parcel into his surrounding larger block. This larger farmer is Attila, the employer we met in the first section. He also has the option of participating in the environmental scheme to preserve his biodiversity, but elects not to, as in his estimation the administrative burden and oversight demanded by the state is not compensated by the subsidy. He is entirely mechanized, with four tractors, and through efforts to consolidate separate parcels of land over the past ten years, with the help of agri-environment and area-based subsidies, he is now able to sow 80-hectare blocks of land with a single crop, and be more selective about which subsidies to apply for. In essence, he has made use of the subsidies in the past to be able to outgrow his 'green' subsidy dependency. Miki, the license-less forester and driver who works for Attila, inherited a single parcel of land, but aspires to consolidate and grow, to mechanize, buy two tractors, and become a smaller version of Attila's landholding, as in his words, 'Our farming is itsy-bitsy, everything takes too long. The only way forward is through the subsidies'. He looks forward to the day that he does not have to await the subsidy's disbursement date expectantly.

These personalities show both a strategic engagement with the state and a dialectical dependency of simultaneous requirement and resentment. When the experiences and aspirations of young farmers are taken into account, a non-mechanized semi-subsistence lifestyle is one many seek to leave behind, either to grow and consolidate their businesses, to sell to larger farm-holders, or to leave. Trouser farming is not an economically viable form of agriculture where markets and flows are international in scale. In the case of the Cserehát, disadvantaged groups bear the responsibility and the burden to maintain environmental goods, shouldering, I would argue, an inequitable distribution of socio-economic disadvantage.

Environmental marginalities

This chapter unpacks the processes and grounded decision-making frames that constitute and rearticulate rural marginalities through three separate but interlinked groupings of people, whose limited opportunities lead to colourful formal and informal attempts to 'get ahead'. These strategies are undertaken by the relatively wealthy and poor alike. Barriers to legitimate work result in flourishing informal economies, without the reward of 'getting rich' or even 'getting by' (cf. Verhage and Shapland, 2013). Indeed, over the years I have heard many variations of the phrase *törvenytisztelet itt abszolút nonszensz* – 'obeying the law here is absolute nonsense', as ability to pay the state-required thresholds in tax or requirements for participation in subsidy programmes are not met. Informal practices are so common as to have a well-known and worn 'lingo', turns of phrase recognizable to those dependent on the same strategies for their incomes and labour, such as *okosba megoldjuk*, 'we'll solve it smartly'. In this light, the objectives of rural development policy to kindle entrepreneurial spirit reflect a lack of grounded understandings of the cultural and economic logics present in isolated, economically poor places, underpinning the constraints and opportunities within which people live in villages in northeast Hungary.

Socio-political history and its shadow have continued roles in the practices and prejudices of countryside populations. Although there continues to be a plural economy wherein 'informal and formal march hand in hand', inhabitants share hardly any of the capacities that define western European 'rurban' modernity, of 'more portable skills, hi-tech communications, portfolio employment and global travel' (Verhage and Shapland, 2013, p. 389). This is an outcome of economic marginality, to be sure: the result of maintaining a lived and geographic distance between the state and people within which economic disparities are only growing, with long-term consequences for state regulatory ability. The persistent urban-rural dichotomy that results in the 'capital and the rest' in Hungary underpins the felt distances between rural lives and their expectations and the sites emanating power, responsible for welfare and subsidy architectures. The geography of rural places does give rise to more than a physical isolation – 'marginality' can here refer to a state that is constituted of the active and passive, chosen and negotiated lived differences and processes in the socio-economic realm, to signify individual and collective 'non-mainstream' choices. The centrality of this meaningful *choice* must never be lost sight of, as discussions on marginality imbue a passive negative positionality that risks not accurately portraying the ambitions of the 'marginal'.

The expectation of uniform application of laws to different socio-ecological contexts becomes problematic in a place like the Cserehát, as it may exacerbate a non-ideal social form of poverty (such as semi-subsistence trouser parcel farming) as well as constrain alternative land uses that may be preferable for communities (as well as offer alternative biodiversity value), such as through afforestation of land areas. Understanding the grounded aspirations and expectations of young men, gypsy workers, and farmers also forces us to question the assumptions of state welfare (whether for living allowance or farming) and whether they make possible or even wanted a movement out of an abject poverty or 'undesired state', as defined by the men in question.

Acknowledgements

I am greatly indebted to János Gyüre, with whom many of the experiences, observations, and their meanings in this chapter were lived and richly debated. Second, to Tatiana Thieme, who has inspired a radical rethink in how to approach, capture, and begin to write down localized informal 'survival strategies', and who approaches her own work in a forever empathetic way.

Notes

1 I acknowledge the significant lack of women and women's places in the Hungarian countryside in this account. However, in my years researching the farming sector in this area I have been without exception the only woman present in the forests, out in the fields, or on the farms during work hours. Women continue to have a predominantly house-bound role in the countryside economy, which certainly merits a gendered, holistic review of younger women's and men's lives in the Cserehát.

2 I use the word 'gypsy' as these workers use the Hungarian *cigány* equivalent when referring to themselves or their communities and not 'Roma'.

3 Translations are my own.

4 One of the key driving rationales for this programme by the majority *Fidesz* government who instigated it is to 'make gypsies work', to break the 'dependency' of local communities on the state, as traditionally (so the political rhetoric goes), gypsies have become too lazy to work and 'milk' the system in a calculating, targeted way.

5 In 2015 the minimum wage was 47,000 HUF per month (US$167).

References

Boels, D., 2013. The Belgian informal economy: a case study of seasonal work in fruit-growing in South Limburg. In: Saitta, P., Shapland, J. and Verhage, A., eds., *Getting by or getting rich? The formal, informal and criminal economy in a globalised world*. The Hague, Netherlands: Eleven International Publishing, pp. 201–224.

Braverman, I., Blomley, N., Delaney, D. and Kedar, A., 2014. Introduction: the expanding places of law. In: Braverman, I., Blomley, N., Delaney, D. and Kedar, A., eds., *The expanding spaces of law: a timely legal geography*. Stanford: Stanford Law Books, pp. 1–29.

Castree, N., 2010. Crisis, continuity and change: neoliberalism, the left and the future of capitalism. *Antipode*, 41, pp. 185–213.

Centeno, M. and Portes, A., 2006. The informal economy in the shadow of the state. In: Fernandez-Kelly, P. and Shefner, J., eds., *Out of the shadows: Political action and the informal economy in Latin America*. PA, USA: Pennsylvania State University, pp. 23–48.

EC., 2005. *Council decision on community strategic guidelines for rural development (Programming period 2007–2013)* [online]. Available at: <http://ec.europa.eu/agriculture/rur/leg/com2005_304_en.pdf.> [Accessed 12 Jul. 2015].

Hann, C. and Sárkány, M., 2003. The great transformation in rural Hungary: property, life strategies and living standards. In: Hann, C., ed., *Post-socialist agrarian question: property relations and the rural*. Munster: Lit Verlag Munster, pp. 117–141.

Hír24, 2014. *Kétszázezer felett a közmunkások száma – Hír24* [online]. Available at: <http://m.hir24.hu/hirek/belfold/Ketszazezer-felett-a-kozmunkasok-szama> [Accessed 14 Jul. 2015].

Kleijn, D., Baquero, R.A., Clough, Y., Diaz, M., Esteban, J. de, Fernández, F., Gabriel, D., Herzog, F., Holzschuh, A., Jöhl, R., R., Knop, E., Kruess, A., Marshall, E.J.P., Steffan-Dewenter, I., Tscharntke, T., Verhulst, J., West, T.M. and Yela, J.L., 2006. Mixed biodiversity benefits of agri-environment schemes in five European countries. *Ecology Letters*, 9(3), pp. 243–254.

Kovács, E.K., 2014. *Farming for subsidies: lived realities of agri-environment in Hungary*. PhD. The University of Cambridge.

Kovács, E.K., 2015. Surveillance and state-making through EU agricultural policy in Hungary. *Geoforum*, 64, pp. 168–181.

Lowe, P. and Ward, N., 2009. England's rural futures: A socio-geographical approach to scenarios analysis. *Regional Studies*, 43(10), pp. 1319–1332.

MTI, 2011. *Pintér: a közmunkaprogram nem cél* [online]. hvg.hu. Available at: <http://hvg.hu/itthon/20111001_pinter_kozmunkaprogram> [Accessed 14 Jul. 2015].

Murdoch, J., Lowe, P., Ward, N. and Marsden, T., 2005. *The differentiated countryside*. London: Routledge.

North, D. and Smallbone, D., 2006. Developing entrepreneurship and enterprise in Europe's peripheral rural areas: Some issues facing policy-makers. *European Planning Studies*, 14(1), pp. 41–60.

Pruitt, L., 2014. The rural lawscape: Space tames law tames space. In: Braverman, I., Blomley, N., Delaney, D. and Kedar, A., eds., *The expanding places of law: a timely legal geography*. Stanford: Stanford Law Books, pp. 190–214.

Saitta, P., 2013. Just a matter of order? An introduction to the topic of informal economy. In: Saitta, P., Shapland, J. and Verhage, A., eds., *Getting by or getting rich? The formal, informal and criminal economy in a globalised world*. The Hague, Netherlands: Eleven International Publishing, pp. 1–10.

Verhage, A. and Shapland, J., 2013. Getting by or getting rich, or simply looking for a life? The informal economy in times of economic crisis. In: Saitta, P., Shapland, J. and Verhage, A., eds., *Getting by or getting rich? The formal, informal and criminal economy in a globalized world*. The Hague, Netherlands: Eleven International Publishing, pp. 389–404.

Woods, M., 2012. Rural geography III Rural futures and the future of rural geography. *Progress in Human Geography*, 36(1), pp. 125–134.

12 Marginality as resource? From Roma people territorial practices, an epistemological reframing of urban marginality

Elisabetta Rosa

Introduction[1]

In November 2013, the Municipality of Turin, historically left-wing, issued an invitation to tender for the 'Implementation of actions in favour of the Roma population'. The document contained an overall strategy to 'overcome the critical aspects of [Roma] settlements, both the authorised and the non-authorised ones' (Comune di Torino, 2013a, p. 4). The contract was finally assigned to the project 'The possible city', proposed by a group of six local associations, some of which had previously worked with Roma people in marginal conditions. According to the tender, the biggest illegal Roma settlement, called 'Lungo Stura' and located on the edges of the Stura River, where nearly 800 Romanian Roma lived, was supposed to disappear. With this aim, the project foresaw the social and housing integration of the people living there. This integration included the rehousing into the private rental market or into self-renovated farms in Turin and its surroundings, and also an assisted return to Romania. In the latter case, Romanian local associations would take charge of the reintegration of the families who went back. The total public funding for the project was 5,000,000 euros.[2] On 26 February 2015, an unexpected eviction took place in the Lungo Stura settlement. Two hundred people were forced to leave by the police just before the bulldozer came to destroy their shacks and caravans: no resistance was put up to the eviction, either by the Roma living there or by local associations.

In the city masterplan (dated 1995 and updated in 2014), the edges of the Stura are represented as an empty space destined to become a river park (Figure 12.1). Waiting for this transformation, due to the flood risk, traffic, parking, dwelling, tents, and shacks are not allowed along the banks of the river.

The film *City Veins – A no man's land/A European suburb/A crossroad for stories*[3] was filmed between 2008 and 2010 on the edges of the Stura. It tells four stories describing what is called 'a no man's land' in a provocative manner: as the film shows, life is flowing in and through these city veins, both in a real and metaphorical sense. Drug dealers and consumers are on one side of the river, and, on the other side, are a homeless person who arrived having lost his job and his home, informal vegetable gardens and the farmers who have tended them for 30 years, and the illegal Roma settlement. They are nearly invisible – from the

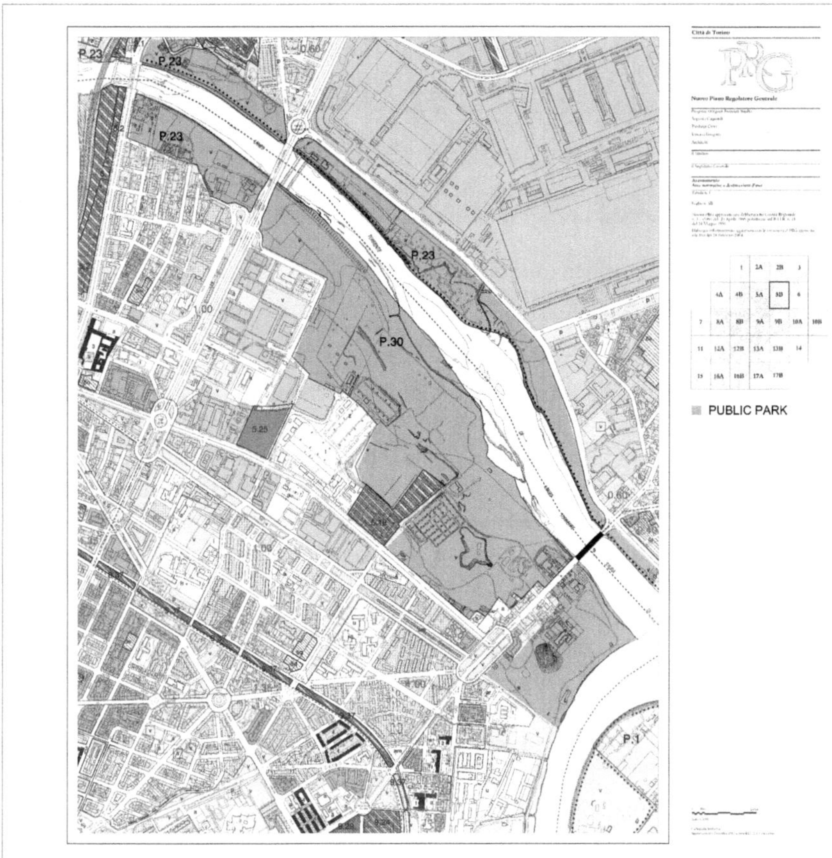

Figure 12.1 The edges of the Stura in Turin masterplan.

Source: Elaboration by Elisabetta Rosa from Comune di Torino, 2014.

bridge, from the road – and nearly invisible interactions exist among them; yet they do exist and interact with the city, and life overflows these supposed physical, sensorial margins (Figure 12.2).

These elements are all focused on the area around the Stura River located at the northeastern periphery of Turin. They differ in the way they look at it: something that has to disappear, an empty space that has to be filled in to define its existence, a vein that contributes to nourishing the city despite its supposed invisibility.

The two Roma illegal settlements that are the object of this research, Lungo Stura and Germagnano, are both located in this part of the city, on the same edge of the river approximately 2 km apart. In Lungo Stura there were nearly 800 people before the eviction while, in Germagnano, there are nearly 250 (Comune di Torino, 2103b).[4] They are almost all Roma people coming from eastern European regions, primarily Romania (henceforth called Roma migrants), and they

Figure 12.2 Stura River seen from the Amedeo VIII Bridge. Panoramica.
Source: Elisabetta Rosa, 2013.

moved to western European cities mainly for economic reasons (Marušiakova and Popov, 2008). They started coming to Turin in the late 1990s (Revelli, 1999), and the main inflows happened during the second half of the 2000s, particularly after 2007 when Romania joined the EU. The two settlements have existed since 2005–2006.

The territory of the Stura has first of all an urban character: it is an expression of the city, rather than of nature, notwithstanding environmentalist rhetoric and regulatory planning tools (Bianchetti, 2011). The urban fabric is fragmented and uneven, composed of multiple functions and spaces: large industrial and commercial plants with their vast car parks, low- and middle-class residential neighbourhoods, urban highways connecting to the motorways to Milan, and stratified and meandering travel networks. Legal and illegal spatial practices and uses; rights acquired, claimed, or lost around private or informal vegetable gardens; abandoned spaces occupied by makeshift Roma settlements: this results in a contested and shared territory where multiple urban dimensions interact, encounter, confront, clash, or simply coexist.

This chapter focuses on urban marginality and aims to reexamine the canonical interpretation of where *the* margin is analyzed and managed, from an academic and political point of view, as an object opposite of the city. Starting from 'marginal practices', of marginal people, in marginal spaces and beyond, I will configure an epistemological reassessment of the notion of marginality (Bradatan and Craiutu, 2012), reframing it as a repository of practice-based resources. I discuss the theoretical assumptions that are the basis of this hypothesis in the first section, where I explain how looking at marginality as a resource could open up new interpretative perspectives. The second part is dedicated to the case of Roma migrants living in illegal settlements in Turin, on the edges of the Stura River. It particularly analyzes spatial practices within and throughout the city. Relying on a case study, in the third part I articulate marginality as a spatial, social, and economic resource, both for marginal people and for other urban actors. The resource lens allows us to overcome the opposition between the city and its margins, to show the porous character of marginality, reveal little known aspects of marginalization processes and the way they are constantly challenged. Finally, I will discuss the political relevance of these implications.

Can marginality be considered a resource? Elements of an epistemological review of marginality

The forced eviction of the Lungo Stura settlement apparently seems to contradict the social and housing integration of Roma people foreseen by 'The Possible City' project. However, I argue that they are expressions of the same interpretation that understands marginality in a dualistic way, stressing the opposition and the dichotomy between margin and non-margin, dominants and subalterns. It is precisely this kind of assumption that I will critically reexamine, as it seems to me to be the main limitation of current approaches to urban marginality. The reductionist assumptions, that is, the attempt to hush up one side of the dichotomy (Derrida, 1972), are one of the main consequences, as it is also the reification of marginality, often described with a single and stigmatizing term: *the* bidonville, *the* slum, *the* camp (Depaule, 2006). Segregation, eviction, the integration of urban margins (the way public policies deal with Roma migrants in precarious conditions is an interesting example of this) are the reflection of these interpretations where the relational character of marginality is hidden, nearly denied, in favour of the hardness of the limit, the edge, the border (Prost, 2004). The relational opportunities opened in, by, or through the margins are consequently forgotten, which would not happen if they were to be considered as a 'trait' that holds together opposite sides (Derrida, 1978). The margin would then become an in-between dimension, that could be understood from the multiple and complex interactions with other spaces around it. *City Veins* is an evocative image that goes in this direction, looking at the Stura's edges from a different perspective, where different people live and interact with each other and with the rest of the city (see also Lancione, 2010).

Relying on critical urban theory (Brenner, 2009; Marcuse, 2009; Brenner, Madden, and Wachsmuth, 2011), I have therefore come to believe in the necessity of a different approach to urban marginality, to which the increasing contamination between marginal and non-marginal, or rather among different urban dimensions, is central. Following Agier (1999) and other authors who highlight the resources that people in marginal conditions are able to mobilize (social relations and networks, social capital) and the various kinds of interdependence that they activate with other social groups (Simone, 2007; Roy, 2009; McFarlane, 2012), I will explore the epistemological value of marginality as resource hypothesis. What else can be learned about the relationship between marginality and the city by considering marginality as a resource? Which kind of resource is it, and for whom?

Two aspects of resources are of particular relevance in this framework: social construction and relative character. Resources do not exist *per se*, that is to say, before or without people qualification, desire, and mobilization (Kahn, 2007; Moroni, 2013). Every object, whether it is material or immaterial – a piece of knowledge, a capacity, a representation – has the potential to become a resource. Its activation depends on the value people give to it and their intentionality. The resource-lens can therefore reveal invisible or unknown aspects of marginality, both because they are hidden in a physical or metaphorical sense, and because they are not recognized as such. Moreover, the relative character of resources (Kebir and Crevoisier, 2008) means that the same object that was once perceived as a constraint or a weakness can discover its potentialities in another spatial and temporal context and become a resource. Both marginality and resources are relative in time and are dependent on context. They both are either possible or actual, depending on the people agency.

Roma people are often represented as catalyst of socio-spatial marginalization and are almost always analyzed within their marginal dimension, that is, the state of social and spatial exception where public policies relegate them (Agamben, 2003). Very few studies on this topic pay proper attention to their agency within the city, the interactions they develop with it, and the multiple resources they are able to mobilize (see, for example, the Urba-Rom Network's approach; also Legros and Olivera, 2014). In the context of this research, Roma migrants' territorial practices are assumed as revealing interactions between the multiple dimensions of the city, throughout their mobilization of resources. As McFarlane points out, practices 'have no pre-given geography or political context. . . . They co-constitute and dissolve spaces' (2012, p. 105). Following this perspective, the analysis concentrated on Roma migrants' practices of using, staying in, and crossing the city, including marginal, non-marginal, and in-between spaces. These practices have been chosen as indicators of more complex activities related to urban resources: mobilization, appropriation, and ordinary resistance (Scott, 1989).

The investigation was conducted by means of an ethnographical approach in 2013 (summer) and 2014 (spring/autumn). It included observation, carried out while following people in their daily activities, and non-structured interviews, to further reveal the development of their practices and views. Both observation and interviews were conducted in different places in the city: in their living spaces (the settlements and their surroundings); in their daily movements (by bus and bicycle, on foot); in their daily activities (begging, waste recovering, selling collected materials at the flea market, washing car windows at traffic lights); and in leisure moments (the marketplace, public showers). At the same time, to further articulate the multiple and relative character of marginality as resource, I analyzed local planning tools, policies, and transformations concerning the area of the Stura. I also conducted semi-structured and non-structured interviews with people interacting with Roma migrants 'around' their living places and within the city: representatives of local associations working in the settlements, representatives of local institutions responsible for social and welfare policies, residents, and local traders.

How marginality is continuously challenged: Roma migrants' practices through the city

Water

The Stura's water has multiple functions. Roma migrants use it for washing dishes and clothes. Children usually go and play in the river; it serves as a swimming pool during summertime, and sometimes they also go fishing. At the same time, the riverbanks serve as a toilet, as water helps to remove waste. The surrounding vegetation helps to protect these private practices from external views.

Public fountains are the main source of drinking water for Roma migrants. Small green fountains with the typical bull's head that spouts water from the nose are a symbol of Turin and they are spread all over the city. They allow everyone free access to water, including Roma and other people in marginal condition (the homeless, squatters, etc.) who use them not only for drinking purposes, but also for washing clothes or dishes. The average distance a person living in Lungo Stura or Germagnano has to walk to reach a fountain is about 500 metres. At least once a day, they collect water with big cans carried in old strollers or hung on a bicycle luggage holder. This task is typically carried out by boys or women, but sometimes men also deal with it. The same water is also used for cooking.

In addition, if we followed the public water geography and Roma practices within the city, we would come to one of the two public shower facilities in Turin. As I could observe, this is not just a matter of having a shower. It is also a rest time to spend with the family, for one or two hours, including a long hair brushing for women and girls and a hot chocolate at the drinks machine for children, while listening to music from their mobile phones.

Waste

Waste recovery is one of the most common activities of Roma people in Turin. Recovered materials include different kinds of objects: metallic ones (kitchenware, household items, spare parts for bicycles), clothes and shoes, food, house furniture, electronic equipment and devices, and books and comics. This activity, which entails the sorting and selling of selected objects in formal and informal economic networks, has multiple spatial dimensions. Most of the people I met usually looked for raw materials, sifting commercial or residential garbage in different parts of the city, depending on the distance from the settlement, looking for the wealthy neighbourhoods or the most populated ones, avoiding those areas where waste collection is door to door (as there are no bins in the street). In any case, waste collecting draws in a deep knowledge of the city, its functioning, its spaces and time, which is needed for a better organization of their work: the schedule of the official collection service so they can arrive before the bins are emptied, and the closing time of marketplaces, shops, and shopping malls so they can get there when the leftovers of the day are thrown out.

Through this waste recovery, Roma migrants use and cross different city spaces at various times of the day and night. Their practice of moving within and throughout the city reveal a sort of 'inverse city': working at night before garbage bins are emptied; hanging out in shopping malls and supermarkets after closing time and waiting for the unloading door of the building (where waste is collected); and meeting in marketplaces when the market is closing. For example, at the butcher at Porta Palazzo, the main city marketplace, I observed Roma women waiting outside the door of the shop. At closing time, the butcher comes to the exit with a large plastic bin with meat remnants. While the trucks collecting food waste are already at work, Roma women select some meat and then throw the rest in the bin. This happens almost every day. This inversion borders on paradox: on the one hand, waste collection and recovery give Roma a certain access to the city and its economic and relational resources, but, on the other hand, the stigmatization coming from this activity is a cause of exclusion and relegation.

Wheels

Whether going to the public shower, carrying out waste recovery, or selling at the flea market, Roma people practice in the city far beyond their living spaces. I had the opportunity to notice that their displacements within the city are very often marked by 'moving on wheels': buses, bicycles, strollers, and trolleys. Particularly, bicycles, strollers, and trolleys are used for waste recovery, depending also on the gender – for example, it is very uncommon to see a Roma woman riding a bicycle. Here again, we come across an 'inverse' way of crossing the city, as most of the people using a bicycle do not care about either going the wrong way down a road or using the pavement instead of the road.

Moreover, the Roma people I met show a deep knowledge of the public transport network, which has become a basic reference point for their use of the city. Instead of street names and civic numbers, bus numbers and bus stops are used in their discourse to provide information about a certain place (a bar, a hospital, a school, the marketplace) or the people living in it. For instance, I often heard people of Lungo Stura talking about 'those of 51' when referring to Roma living in Germagnano, from the number of the only bus that stops nearby.

Sometimes it happens that they get lost (quite frequently when they go for garbage, as they are only focused on collecting as much as possible). Then, the bus network is their safety net; they learn the bus numbers they have to get by heart and they repeat them as a *ritournelle* (Deleuze and Guattari, 1980) until they find the right way back. This is, for instance, the case of Marius, a 60-year-old man I met begging in front of a supermarket. He hardly speaks Italian; the only clear information I could get from him was, 'Tramway . . . 18, 27, 49 . . . Tramway . . . 18, 27, 49'.

Pavements

Many Roma I met were begging. There is no room here to deeply analyze begging itself and the different ways it could be practiced (Dean, 1999), yet what is worth considering is its spatial articulations and the resulting social interactions. Begging could be either a static or a moving practice, but, in the case of Roma migrants, on the basis of what I could see, it is almost exerted in a punctual way, sitting on the pavement at the entrance of hospitals, shops, churches, supermarkets, and shopping malls, or standing at traffic lights. Several factors contribute to the choice of place for begging, but the basic element is people interaction, either in terms of flows (the number of people likely to be met) or in the kind of interaction (a good relationship with the store owner or residents of the neighbourhood). A long-lasting presence on the same corner of the pavement, the same step, or at the same crossroads lets people get used to the beggar and increases trusted relationships. This is for instance the case of Stefania (40 years old). Every Sunday she comes to the church of the Consolata to beg. Her 10-year-old son comes with her and waits for her in the bar next to the church while she works. There is a television in the bar (they don't have one in the settlement) and the boy is happy to spend the morning watching cartoons. The bar managers have looked after him every Sunday for five years. Coming to the same pavement every day for years can also be considered a form of space appropriation or even a kind of ordinary resistance.

What is interesting to highlight is also that the places the beggars occupy are often marked by their presence even when they are not there, as they normally leave a pull or a piece of cardboard on which they sit, a bottle of water, or other kinds of objects they use for this activity. The encounter of these materialities – the pavement and the pull – configures a sort of absence-appropriation of spaces and informs us about the interactions between the people and the city.

Moorings

Roma migrant living spaces in Lungo Stura and Germagnano are built 'individually and incrementally, using locally available materials, and often clustered in ways that depend on closely shared roofs, walls and infrastructures' (McFarlane, 2011, p. 216). Different kinds of materialities shape them and inform us simultaneously about both their ephemeral and solid character. Homes are very often composed of two or three housing units (shacks or caravans) with a shared roof; the floor is made of concrete or tiles; shacks are decorated with flower pots outside the window or at the entrance door, and the interior space is decorated with tapestry, paintings, photos, and fake flowers. Plexiglas or clear plastic is used for windows, even if sometimes a whole window frame can be found; corrugated iron for roofs, often covered with plastic to make them waterproof; wooden boards are assembled to build internal and external walls and doors; concrete or tiling to isolate the ground and keep it clean. A carpet can be found at the entrance door, which is also marked by a wooden or plastic board vertically placed to prevent rats from entering.

Housing is not the only function of the settlements; bars, places of worship, and a shack for educational and leisure activities led by various associations can also be found. Dense vegetation surrounds the settlements, ensuring protection against inclement weather and public exposure. As the vegetation becomes less dense, small open spaces are organized for waste recovery, such as car and bicycle repair, separation of copper from other metals, and sorting recycled material to then sell it within the settlement or at the main city flea market. On a closer look, all of those materialities used by Roma during their daily practices can be found everywhere in the settlement: bicycles, trolleys, and strollers on rooftops to prevent plastic flying away, or maintaining barracks walls, and empty plastic cans used for sitting. As McFarlane argued, within these processes of assembling and reassembling materials, 'the city is both *mined* and *recycled*. . . . Housing is, in short, both *made* and *edited*, in context of deeply unequal resources and precarious lives' (2011, p. 216).

Marginality as resource: towards an epistemological reassessment of the notion of marginality

Marginality as resource has an epistemological value, serving here to explore how marginality is lived, managed, challenged, and reaffirmed in the various spatial and temporal contexts of the city. Relying on the analytical framework outlined above, I will reframe marginality as a repository of practice-based resource and articulate it as a spatial, social, and economic resource.

Spatial resources

Natural, transforming, and abandoned areas – these are all different spaces that nobody can apparently use, but could be an important spatial resource for urban transformations or non-transformations, for marginal people and other urban actors.

The environmental and ecological resources of the Stura River basin form the main value that planning tools identify for this territory. The Stura area has a strategic role within the Green Ring (*Tangenziale Verde*) project all around the city, as proposed by inter-municipal planning. As for local planning, the city masterplan foresees the creation of the Urban and River Park of *Basse di Stura* and the Amiat Park, this latter in place of the municipal waste dump closed in 2009. In none of these documents are Roma illegal settlements mentioned.

During the 20th century, many factories were established in the northeastern periphery of Turin, following the railway's development towards Milan and the eastern Italian regions, and attracted by the Stura River. It was used as a source of energy and as a place to dump waste production. Over the years, its basin has become the city's waste tip, hosting all of the activities the city decided to keep distant from the centre: the municipal waste dump, the municipal dog pound, and three authorized settlements for Roma people, called 'nomad camps'.[5] Other non-authorized activities then gradually arrived, with the assumed non-attention of municipal authorities: illegal dumping; informal vegetable gardens; drug markets, and illegal Roma settlements. The 'back side of the city' and its empty spaces are an important resource in which to hide urban waste (Figure 12.3).

Over the last ten years, part of the riverbanks have been gradually cleaned up and recovered, namely the western bank where the drug market was. The Arrivore Park was inaugurated in 2009; it includes municipal vegetable gardens allocated to urban farmers. Until 2004, the same piece of land hosted an authorized Roma settlement where Balkan Roma lived. They were then relocated in a new nomad camp on the opposite side of the river, in via Germagnano, near the municipal dump and dog pound. The illegal settlement of Germagnano arrived soon afterward.

Figure 12.3 The 'back side of the city'.

Source: Elaboration by Elisabetta Rosa from Comune di Torino, 2014.

It is specifically in this superseded space between the past and the future, waiting for the transformation to come, in these territorial asymmetries (Legros and Olivera, 2014) that Roma migrants found the possibility of settling on the Stura's edge. Low land values are, in this case, very helpful: the flood risk makes it impossible to build, so housing and property developers have no interest in investing in this area. The lack of private and public attention makes this a 'no man's land', an important spatial resource for Roma migrants. Water, wood, and vegetation helps to satisfy basic needs and maintain a certain invisibility. There are discount stores and supermarkets in which to do the shopping and at which to beg, public transport and motorways to access the city and to leave it, and public fountains to provide clean water. Urban services and infrastructures are not far and the city is all around.

Under the supposed non-attention of the public, life was being continuously assembled, until the moment when the area became the kind of resource planning policies identify for development. A cycling lane will be realized along the river on the eastern bank in the place of the Lungo Stura settlement and a 'public use' of this area will be granted (see below).

Social resources

Social resources can be articulated through the transactional paradigm (Rémy, Voyé, and Servais, 1978; Rémy, 2005) which has, in this case, a heuristic function (Fusulier and Marquis, 2008). According to this perspective, every social practice is a transactional moment among individual or collective actors that are situated in a spatial and temporal context and affectively engaged in the relationships they have with the surrounding world. The transaction takes place when actors enter a process of resources appropriation, where this appropriation can assume different aspects, such as the examination of available capital, the constitution of a symbolic framework, the connection with another individual.

Both begging and obtaining garbage entail interactions between Roma migrants and other city inhabitants. Local residents who get used to the frequent transit of these waste collectors activate direct relationships with them. Some people I happened to meet told me how they pack unwanted clothes or objects and then leave the pack next to the entrance of their home. They are sure someone will come across it during the morning: 'The Roma guy I see from my window'.[6] They do not throw the pack into the garbage bin to prevent collectors having direct contact with waste, which also means the donor does not take part in this activity. Thus, thrown-away objects are saved from becoming waste and retain a certain dignity which is reflected in how the receiver acquires them. Also local traders get used to beggars and waste collectors and give them the unsold goods, as in the case of Roma women waiting for the butcher at Porta Palazzo marketplace. Other kinds of transactions do not involve consumable goods, but time and care, as shown in the story of Stefania and her son.

Social resources for Roma migrants also include the use of social services provided by local institutions, despite their illegal housing condition and justified by the 'emergency condition' of these populations (Sigona, 2005). Social workers who bring children to school, local parishes that offer meals, associations that organize various kinds of activities (religious, cultural, sporting, recreational): all of these actors transmit information, facilitate service provision, and allow a connection to the social network. They provide support and constitute relationship capital (Manzoni, 2014). Moreover, all of these different actors acknowledge the 'empty space' of the settlement that the planning tools do not recognize.

From an inverse perspective, Roma illegal settlements represent an important resource of negotiation around urban transformation. The Lungo Stura settlement is located on private property. To implement the masterplan, the Municipality of Turin has to acquire the land via a procedure of expropriation for public benefit, balanced with money repayment to the landowners. This kind of procedure often generates conflict, where the problem is to define the cost of public benefit in order to satisfy landowners. In 2011, the Municipality of Turin approved the preliminary project to create a cycling lane along the Stura, including the expropriation plan. The final project was approved in 2014. The long-lasting presence of the Lungo Stura settlement has had a strong impact on the natural areas (lack of sanitation, waste accumulation). Local residents periodically report the smell and the noise coming from the camp to the authorities, including poisonous gas generated by burning waste. In 2008, landowners reported the illegal occupation to the Municipality of Turin, asking for its eviction. The demolition of 60 shacks in February 2015 was the first step to the redesignation of the area. Marginal spaces become reserved spaces (*espaces-réserves*: Benach and Tello, 2014): identifying areas as reserved for urban projects means they are intentionally left degraded and can be squatted upon which can then legitimize the upgrade. Roma presence is not always accidental; it can be used to facilitate the negotiation procedure among landowners, local residents, and public institutions (Rosa, 2015).

Economic resources

Economic resources also need to be mentioned because of Roma participation in formal or informal economic circuits, including the selling of recycled materials at the flea market or commercial activities within the settlement.

Living in the informal settlement can itself be a resource, as it allows inhabitants to reduce rent and save money. As Manzoni noticed, '[I]nvesting money in a home is perceived as a useless cost. Many Roma migrants have left home for overseas and transnational behaviours such as remittances justify a temporary and precarious accommodation in a slum' (2014, p. 3). The choice to live in a shack in Lungo Stura can therefore be interpreted as an 'alternative strategy' to formal housing property and rental market, beyond legality and public control (Aguilera, 2011). De Soto (2000) states in this regard that informality contains the grassroots of poor people rebellion against state bureaucracy; he argues that informality is a sign of heroic entrepreneurship.

Far from the celebration of this 'subaltern urbanism', I join Roy when she says that 'informality must be understood as . . . a logic through which differential spatial value is produced and managed' (2011, p. 233). Indeed, as is the case of Lungo Stura, informal settlements are more often than not under public control. The Nomads Patrol is a special police unit that was created in 1982 expressly to control Roma people settlements, both the authorized and the illegal ones. During my fieldwork I could see their yellow Fiat Panda crossing Lungo Stura and Germagnano up to three times a day. Besides, local authorities used to mark the shacks with a number painted in black on the external walls to control the growth of the settlement. Maintaining informality and precarious living conditions ends up being a resource also and above all for the Municipality and for other urban actors interested in those lands. In other words, marginal spaces are *espaces-réserves* also as they represent an important economic resource linked to land value and valorization throughout urban investments and transformation (Benach and Tello,

2014). These dynamics are not evident in the case of Stura River's edges: due to the impossibility to build on these lands, any investment or real estate project is unlikely to be proposed. However, I argue that similar processes are at stake also in this case, allowing the Municipality to bring down expropriation costs. In this case spatial, social, and economic resources are cross-cutting. When the Municipality decided to implement the masterplan, the eviction of the settlement is a necessary step to this transformation which is justified by the public benefit of the river park and cycling lane and which entails an annihilation of what had filled in this space over time, in addition to an economic advantage.

Conclusion

In this chapter, I critically analyze the case of Roma migrants living in illegal settlements in Turin to show that marginality is not an element *per se* nor a condition that can be labelled *a priori*, as established paradigms often do, nor a static or a definitive one; rather it is an ongoing multifaceted dynamic open to the unpredictable (Lancione, 2014). The various kinds of resources mobilized by different urban actors result in a constant arrangement and rearrangement of the city and of its social and spatial margins. Depending on the way the resource-margin is mobilized and by whom, marginality could reveal those dystopias or heterotopies that otherwise would not be perceived.

Saying that marginality can be interpreted as a resource is not denying the fact that it is a central issue in urban studies and policies. Rather, this actually brings it back in the field and recognizes its epistemological value, which also has significant political relevance. Considering marginality as a resource, the multiple and complex dynamics that contribute to marginalization processes and producing and reproducing socio-spatial inequalities appear to be significant. If a system can be understood from its marginal dimensions and the practices developed (Foucault, 1975), marginality is therefore indicative of a broader urban social and political functioning, its complexity, contradictions, and challenges.

At the same time, through the resource-lens, the opposition between the margin and the city seems to become fainter, even disappearing. Marginal people practices are not limited to the construction of 'their' marginal space; neither are they simply an effect of other actors' decisions. They co-constitute the city in an ongoing process of territorialization and reterritorialization, continuously challenging their supposed marginality. The way Roma migrants use the public transport network is an interesting example. This is not to make an apologia for self-help or to celebrate marginal people's capabilities as a creative strategy of the poor, which would give those practices a moral value rather than an epistemological one (Chatel and Soulet, 2002). This is rather to reveal the agency of marginal people and practices within the city and to recognize them as urban inhabitants, and not only as a reified otherness, a passive subject of public action to be managed through exclusion/inclusion and segregation/integration. They are a constituent part of the city, as are the different spaces they practice and the resources they mobilize.

I want to highlight the political relevance of such recognition, as it can open up important perspectives on different ways to consider and administer spatial and social marginality. As in the case of Roma migrants, this recognition is a basic process to de-label the 'Roma question' and consider it within a larger urban question where the main issues at play are city accessibility and liveability and the non-marginalization of certain people rather than spaces. This is more generally to recognize marginality as a constituent dimension of the city which is, at the same time, porous, ephemeral, temporary, and relative. All of these characters of marginality assure the constant becoming of the city itself.

Against this background, a key challenge is the continuous exploration of the multiple dimensions of marginality, not to make it vanish among the thousands of urban possibilities but to bring it back in and start to differentiate among its various phenomenologies within the urban. This seems to me the most interesting perspective to then overcome the 'unique political agency' (Roy, 2009, p. 235) that characterizes the way with which urban margins are dealt.

Notes

1 This work has been realized in the framework of the Laboratoire d'excellence LabexMed – Les sciences humaines et sociales au cœur de l'interdisciplinarité pour la Méditerranée, with the reference number 10-LABX-0090. This work has received financial support from the State managed by the Agence Nationale de la recherche under the project Investissements d'Avenir A*MIDEX with the reference n°ANR-11-IDEX-0001–02.
2 In 2012, the Municipality and the Prefecture of Turin received 5,000,000 euros from the central government to implement the 'National Strategy for the Inclusion of Roma, Sinti and Caminanti Communities', following the Communication of the European Commission n.173/2011.
3 Directed by Andrea Deaglio, Italy, 2010, 63'.
4 Romanian Roma presence in Turin is currently estimated to be 1,000 to 1,500 people (Comune di Torino, 2013b) and approximately 50,000 in Italy (Cingolani, 2011).
5 *Campi nomadi*: they were introduced by Italian public authorities in the late 1980s and during the 1990s to host Roma people coming from the former Yugoslavia and for other Roma groups already present in Turin (the Sinti).
6 Silvia, fieldwork September 2014.

References

Agamben, G., 2003. *Lo stato di eccezione*. Torino: Bollati Boringhieri.

Agier, M., 1999. *L'invention de la ville. Banlieues, township, invasions et favelas*. Paris: Edition des Archives Contemporaines.

Aguilera, T., 2011. Struggling with illegality against housing crisis in Paris. Squats and Slums as a Challenge to Urban Policies. *Working papers du Programme Villes & Territoire*, 3. Paris: Sciences Po.

Benach, N. and Tello, R., 2014. Des espaces-réserve versus des espaces de résistance? In: B. Florin, O. Legros, N. Semmoud and F. Troin, eds., *Marges urbaines et néolibéralisme en Méditerranée*. Tours: Presses Universitaires François Rabelais, pp. 43–57.

Bianchetti, C., 2011. *Il novecento è davvero finito*. Roma: Donzelli.

Bradatan, C. and Craiutu, A., 2012. Introduction: The paradoxes of marginality. *The European Legacy: Towards New Paradigms*, 17(6), pp. 721–729.

Brenner, N., 2009. What is critical urban theory? *City*, 13(2–3), pp. 198–207.

Brenner, N., Madden, D.J. and Wachsmuth, D., 2011. Assemblage urbanism and the challenge of critical urban theory. *City*, 15(2), pp. 225–240.

Chatel, V. and Soulet, M.-H. eds., 2002. *Faire face et s'en sortir. Volume 2: Développement des compétences et action collective*. Fribourg: Editions Universitaires Fribourg Suisse.

Cingolani, P. ed., 2011. *(Rom)eni tra Italia e territori di partenza. Vita quotidiana, rappresentazioni e politiche pubbliche*. Torino: FIERI.

Comune di Torino, 2013a. *Capitolato Speciale d'Appalto per la gestione di iniziative a favore della popolazione rom*. Bando di Gara 84/2013.

Comune di Torino, 2013b. *Rapporto 2013* dell'Osservatorio Interistituzionale sugli Stranieri in Provincia di Torino.

Dean, H. ed., 1999. *Begging questions. Street-level economic activity and social policy failure*. Bristol: The Policy Press.

Deleuze, G. and Guattari, F., 2013[1980]. *Mille Plateaux. Capitalisme et schizophrénie 2*. Paris: Les Éditions de Minuit.

Depaule, J.C. ed., 2006. *Les mots de la stigmatisation urbaine*. Paris: Éditions de la MSH.

Derrida, J., 1972. *Positions*. Paris: Minuit.

Derrida, J., 1978. *La vérité en peinture*. Paris: Flammarion.

De Soto, H., 2000. *The mystery of capital: why capitalism triumphs in the West and fails everywhere else*. New York: Basic Books.

Foucault, M., 1975. *Surveiller et punir. Naissance de la prison*. Paris: Gallimard.

Fusulier, B. and Marquis, N., 2008. La notion de transaction sociale à l'épreuve du temps. *Recherches sociologiques et anthropologiques*, 2, pp. 165–178.

Kahn, R., 2007. Une approche pluridisciplinaire de la dimension culturelle du développement territorial. *XLIIIè colloque de l'ASRDLF «Les dynamiques territoriales débats et enjeux entre les différentes approches pluridisciplinaire»*. Grenoble-Chambéry 11–13 July 2007.

Kebir, L. and Crevoisier, O., 2008. Cultural resources and regional development. In: P. Cooke and L. Lazzaretti, eds., *Creative cities, cultural clusters and local economic development*. Cheltenham: Edward Elgar, pp. 48–69.

Lancione, M., 2010. Giustizia sociale, spazio e città. Un approccio teorico-metodologico applicato a un caso studio. *Rivista Geografica Italiana*, 117(3), pp. 625–652.

Lancione, M., 2014. Assemblages of care and the analysis of public policies on homelessness in Turin, Italy. *City*, 18(1), pp. 25–40.

Legros, O. and Olivera, M., 2014. La gouvernance métropolitaine à l'épreuve de la mobilité contrainte des «Roms migrants» en région parisienne. *EspacesTemps.net* [online]. Available at <http://www.espacestemps.net/articles/la-gouvernance-metropolitaine-a-lepreuve-de-la-mobilite-contrainte-des-roms-migrants-en-region-parisienne/> [Accessed 21 May 2015].

Manzoni, C., 2014. Why Roma migrants leave or remain in settlements? Different strategies of survival and adaptation among Roma in Northern Italy. *2nd International Congress on Resilience: from person to society*. University of West Romania, Timisoara, Romania 8–10 May 2014.

Marcuse, P., 2009. From critical urban theory to the right to the city. *City*, 13(2–3), pp. 185–197.

Marušiakova, E. and Popov, V., 2008. Les migrations des Roms balkaniques en Europe occidentale: mobilités passes et présents. *Balkanologie*, 9(1–2) [online] Available at <http://balkanologie.revues.org/972> [Accessed 21 May 2015].

McFarlane, C., 2011. Assemblage and critical urbanism. *City*, 15(2), pp. 201–224.

McFarlane, C., 2012. Rethinking informality: Politics, crisis, and the city. *City*, 13(1), pp. 89–108.

Moroni, S., 2013. Afterword: Ethical problems of contemporary cities. In: C. Basta and S. Moroni, eds. *Ethics, design and planning of the built environment*. London and New York: Springer, pp. 197–212.

Prost, B., 2004. Marge et dynamique territorial. *Géocarrefour*, 79(2), pp. 175–182.

Rémy, J., 2005. Négociations et transaction sociale. *Négociations*, 3, pp. 85–95.

Rémy, J., Voyé, L. and Servais, E., 1978. *Produire ou reproduire? Une sociologie de la vie quotidienne*. Brussels: Les Editions Vie Ouvrière.

Revelli, M., 1999. *Fuori Luogo. Cronaca da un campo Rom*. Torino: Bollati Boringhieri.

Rosa, E., 2015. Aménagement urbain et marginalisation socio-spatiale. L'exemple des migrants roms à Turin. *Les Cahiers d'EMAM* [online]. Available at <http://emam.revues.org/1101> [Accessed 6 January 2016].

Roy, A., 2009. Strangely familiar: Planning and the worlds of insurgence and informality. *Planning Theory*, 8(1), pp. 7–11.

Scott, J.C., 1989. Everyday forms of resistance. *The Copenhagen Journal of Asian Studies*, 4, pp. 33–62.

Sigona, N., 2005. Locating "The Gipsy Problem". The Roma in Italy: Stereotyping, labelling and the "Nomad Camps". *Journal of Ethnic and Migration Studies*, 31(4), pp. 741–756.

Simone, A., 2007. At the frontiers of the urban periphery. *Sarai Readers 07: Frontiers*, pp. 462–470 [online]. Available at <http://archive.sarai.net/files/original/e336645 1a06d9bbe4187fac6a550ec91.pdf> [Accessed 6 July 2015].

13 Citizen participation as microfascism

Marginalizing labour in Web 2.0

Cheryl Gilge

The creative environment of Web 2.0, powered by user participation, democratizes production and gives agency to average citizens by connecting them to their community. The citizen-*fill-in-the-blank* embraces these roles as citizen cartographers, citizen scientists, and citizen journalists, using crowdsourcing platforms and social media. This production energizes hybrid research methods and creates vast stores of data as an object of study. *And* a marketable product to sell. As the digital economy of 2.0 challenges traditional business models, it shifts labour relations and creates new attendant practices. As more corporations, researchers, and governments look to this 'lean production' to capitalize on the efficiency of technology and the enthusiasm with which users adopt it, the *value* of labour is marginalized. Wage-selling actors compete with the free labour of willing participants.

The desire to participate, the creative potential, and the perceived need for visibility all offer a view onto the tension that exists within the 2.0 environment. This unique intersection for collaboration and self-inflicted marginalization of one's labour stands as a contemporary example of 'desiring our own repression'. It oscillates between willful volunteerism and what Deleuze and Guattari (D&G) call microfascisms: daily practices adopted at the expense of individual freedom, with a belief in a better future or certainty of outcome. This chapter proposes that 'microfascism' is a robust analytic to better understand the tensions that exist between social production *and* individual participation, as an investment of desire to be a part of a larger goal. Herein, this chapter sketches out the current Web 2.0 environment, the valuing or marginalizing labour in this new digital economy, and whether rational motivation or desire animates the citizen producer. Finally, it identifies moments in which this 'investment' turns from empowerment to enslavement, highlighting the blurry boundaries between Web 2.0 as a radical, democratic alternative to capitalism and the potential emergence of cultural fascism.

The Web 2.0 milieu

Web 2.0 serves as an umbrella term for the current Internet 'ecosystem', which signifies the read-write capacities of software architectures that are reshaping contemporary life (Dijck, 2013). Here I focus on the three distinct practices that largely define the Web 2.0 milieu. In the first instance, open source is a libertarian response and collective effort to produce an alternative to corporate software and leverages individual contribution within a larger project. Second, Web 2.0 architecture democratizes individual contribution, whereby users become producers of online content. Finally, crowdsourcing harnesses the potential of individual production, drawing on crowd diversity and participatory spirit. Taken together, the *ability* to contribute and *willingness* to receive contributions defines the characteristics of the 2.0 culture; horizontal, 'open', and democratic defines the particular spirit.

There are two key shifts important to the emergence of the present Web 2.0 culture. First, the open source movement emerges as a challenge to corporate software with its early success of Linux. Legions of developers donate labour time, forming a community that is defined by code's technical rationality and belief that software should be shared (Weber, 2004). The ethos that guides this community shares an enthusiasm for and evangelizes technology as a critical tool for efficiency and innovation (Raymond, 1999). Second, the user emerges as an active participant, as sharing and uploading content become effortless. With an increase in sophisticated platforms, this burgeoning social system shapes and creates new subjectivities that capture this spirit and alters daily practices. Co-producer appellations include citizen-*fill-in-the-blank*, prosumer, netizen, etc. Rhetorical appeals actively draw on ideas of sociality and participation (Dijck, 2013) and highlight collaboration and community by invoking creativity, innovation, transparency, or efficiency (Brabham, 2013). Together, this unique environment infuses society with a creative and entrepreneurial spirit, one that upholds ideals of 'freedom' through autonomy and efficiency. It cuts across a range of interests that benefit from Web 2.0 production: lean and horizontal business strategies to access 'fresh' ideas; civic initiatives for government efficiency; as an object of study; and means to extend research dollars.

There are a diversity of platforms and agendas that solicit the productive user. These include creative work of coding for open source projects like Linux; app development for the mobile market (open source and proprietary); contest structures for new products (chip flavours, t-shirt designs, shoes); data and scientific problem solving (InnoCentive, Foldit); self-publishing and promoting everyday activities and witty political commentary (videos, photos, likes, tweets, blogs). While hardly exhaustive categories, all of these activities exist within the same Internet ecosystem. Technology layers onto the intellectual and affective qualities of humanistic modes of communication and social needs and transforms it into surplus value, out of desire for creativity, community, intellectual challenge, autonomy, and efficiency.

Collectively, most users benefit from this open horizontal network through user-friendly platforms; low cost of access enables easy participation, which gives voice to a greater diversity of constituents. Adherents are enthusiastic about the radical, democratic potential of 'freedom of information' for social movements. These modes also directly challenge the proprietary capitalistic logic and business models of profit-driven commodities, as open source software is free to distribute. Empowering communities with volunteer efforts and open source code eschews corporate software. Social media platforms challenge traditional media models of consumption. Amateur production on creative platforms like YouTube and Flickr rivals professional effort, capturing large audiences and notoriety. Peer-to-peer sharing transgresses outmoded property rights, as digital files are infinitely reproducible.

At the same time, prognosticators are breathless about Web 2.0's potential to radically rethink 'business as usual' and create dynamic product development. Tapscott and Williams (2007) coin this environment *wikinomics*, and Howe (2006) describes a new economic trend making foreign 'outsourcing' passé. They both point to the innovative potential unleashed through top-down organization, sentiments of a pro-business economy. Creative 'design thinking' and technology inject lean 'start up' agility into traditional, large corporations, extending this spirit of innovation, creativity, and horizontality in all directions. This enables open business models and closer customer collaboration.[1] Web 2.0's networking capacity enables diffuse connections between workers and employers, expanding the definition of 'freelancer'. Platforms offering earning potential are on the rise, as tech evangelists highlight the flexibility of 'just-in-time' employment (TEDx, 2013). Sectors utilizing this new mode are diverse, and include research, professional services, or editing and transcription, as well as more mundane tasks like household chores and services, delivery, ride shares, etc. It appeals to the entrepreneurial spirit and a 'self-made' mythology. It also opens up new labour markets free from government regulations. This micro task-for-hire strategy shifts both the workday and employment models, as individuals accept tasks during leisure time from any location.

The benefits for business are resoundingly clear. For libertarian persuasions, crowdsourcing and technology can also tackle 'society's most intractable challenges' (Eggers, 2013). This reduces the government's role and increases individual autonomy by connecting them directly with private sector resources. Websites like opensourceforamerica.org and codeforamerica.org are enthusiastic about technology's ability to revolutionize government operations. Researchers and non-profits look for ways to harness this enthusiasm and see great potential when human sensors generate data of on-the-ground conditions or documenting community resources (Goodchild, 2007), and seek ways to exploit its productive capacity for research and learning (Dodge and Kitchin, 2013). Big data and Volunteered Geographic Information (VGI) are new areas of research and Citizen Science's newly formed Association point to the importance and desire to expand research capacity by harnessing crowdsourcing's potential and creatively cutting overhead costs of clipboard 'foot soldiers' (Grabar, 2015).

Valuing (and marginalizing) labour in a digital economy

The new forms of communication, globalized economic relations, modes of production, and the spatial and temporal flexibility of Castells's 'Information Society' now appear quaint in comparison to the steroid-esque environment of Web 2.0. The flexibility and precarity of neoliberal policies are greatly expanded by the horizontal earning platforms. As striving, self-generated production and entrepreneurial spirit become a celebration of freedom, shifting norms make this a standard condition (Lazzarato, 2012), especially for the under-40 demographic (Ross, 2013). The volunteered labour from coding, crowdsourcing, and Web 2.0 production represents a significant expenditure of labour time. How to *value* this labour is less clear, when the lines between leisure and labour are increasingly blurred.

Contemporary labour studies alight on the contradictions stemming from immaterial labour for services and products, which trouble classic Marxist categories of labour and commodities (Huws, 2014). Immaterial labour transforms manual into intellectual labour; this restructures the worker, who is given more autonomous control over his or her work (Lazzarato, 1996). Labour no longer takes place on the proverbial assembly line; new mobile working conditions give 'creatives' more flexible work arrangements in diverse environments. This creates new diffuse management techniques, as mobile technology ensures one is always reachable. This blurs the division between domestic and professional spheres (Ross, 2013): work is distributed across a larger portion of the day and email correspondence is often uncounted hours (Huws, 2014).

Data points and lines of code become the source of value in this digital economy, as autonomous advertising schemes generate the lion's share of revenue. *Cognitive* rent provides a different perspective on valuing this production, given algorithms, code, and data's ability to continue generating value (Pasquinelli, 2009). The labour animating this new immaterial production creates new typologies of worker status, acquiring appellations such as 'precariat' (Standing, 2012) and 'cybertariat' (Huws, 2014). Scholars debate the quality of these asymmetrical power relations, and include exploitation (Fuchs, 2014), self-exploitation (Ross, 2013), alienation (Andrejevic, 2013), free labour (Terranova, 2000), and precarity (Standing, 2012). It is no longer 'dead labour' that resides within the material object; rather, knowledge and experience remain in the possession of the worker and a source of value for future prospects (Ghosh, 1998). With social media platforms, users remain in possession of the thoughts and experiences while generating actual value as an active (captive) audience for advertisers and monetizable data points.

Effortlessly produced and eagerly accepted, *what* is being produced and *where* the value resides remain important questions. Various entities directly and indirectly benefit from this contribution, either as access to free ideas, data, or financial gains that result from this new knowledge 'acquisition'. This free production marginalizes or devalues labour value in two distinct ways. First, prizes or recognition are offered as token of appreciation for participation, rather than monetary pay. Individuals identifying with a product, community, or cause are more likely to contribute (Lindenberg, 2001). The perceived value of the 'token' correlates with an individual's intrinsic or extrinsic needs (Howe, 2008), while the actual costs of the prize structure are below the value of labour hours contributed. Well-known prize-structured projects include Threadless t-shirt design contests, Google's Mapathon initiatives, or Goldcorp's Mining Challenge, although examples are numerous. After failing to locate new gold stores, Goldcorp offered a $500,000 reward for winning solutions. Collectively, the labour hours went unreported, while the mining solutions proffered yielded $3 billion in gold (Brabham, 2013). In Google's India Mapathon, individuals interested in prizes and visibility dramatically outperformed individuals with only a desire to map their community, with performance measured by the quantity of map edits. Individuals contributed spatial knowledge and became part of a mapping community, receiving certificates, t-shirts, and a small number of mobile devices as prizes. Google received a more comprehensive map, and reasserted their market share, as result of *all* map edits.

Moreover, the culture of 'connectivity' of Web 2.0 and social media dynamically reshape ideas of recognition and notoriety, where this new social currency influences daily practices (Dijck, 2013). Open source project participation is overwhelmingly community driven, and diverse intrinsic rewards include skill building, creative outlets, and intellectual challenges (Lakhani and Wolf, 2005), while extrinsic rewards include signalling incentives (Lerner and Tirole, 2002). New models of journalism are changing online readership and creating new avenues of notoriety for aspiring media producers. Corporate media channels increasingly access this content produced by marginally paid or volunteered labour. It enables media to expand its reader base and capture more advertising dollars without increasing employment costs (Howe, 2008). The sale of *Huffington Post* exemplifies the potential value of this labour. Bloggers generated content, received no compensation except 'exposure', and were denied legal recourse to the inordinate financial gain of $315 million that Huffington and her investors ultimately enjoyed (Ross, 2013). The general increase in soliciting reader photos, user-generated stories like iCNN, monitoring social media for trending issues, and extensive reblogging of content extends this production in all directions.

The *proliferation* of projects utilizing Web 2.0 strategies further transforms this marginalization of labour by legitimating its use across sectors in myriad ways. Appealing to the online community via social media is a now *de facto* strategy. 'Like' buttons and clever hashtags are deployed in myriad situations, including contests, generating brand interest, and civic participation in governance. This social activity raises cultural awareness, and the preponderance of data generates monetizable user profiles (Fuchs, 2014). This adoption also transforms a novel crowdsource solution like Goldcorp's mining challenge into a standard practice and one marketing strategy among many. The 'intrapreneurial' company, Board of Innovation, exemplifies this by helping corporations 'innovate like a start up', listing crowdsourcing as an effective business strategy. As a whole, this transformation demonstrates the broad social acceptance of technology platforms (Dijck, 2013) and legitimates additional efforts to harness crowd 'wisdom'. The diversity of volunteers or cheap labour in a Web 2.0 environment changes the dynamics of participation with increased means of attracting participants. Creatively capturing increased contribution often includes using game structures and technology, which is effective at capturing the 'millennial' demographic (Bowser et al., 2013), where even researchers are eager to tap this enthusiasm.

As participants in the larger scheme, business models soliciting the idle producer extend the general environment of precarity. Jeff Howe's famous coinage of crowdsource was a veritable, unapologetic riff: 'Remember outsourcing? Sending jobs to India and China is so 2003. The new pool of cheap labour: everyday people using their spare cycles to create content, solve problems, even do corporate R & D' (2006, p. 176). Amazon's Mechanical Turk popularized the corporate model of small tasks for hire. Since 2008, there has been as much as a 700% increase in micro-task platforms and projects, according to industry reports (Crowdsource.com). As a pro-business model of accessing cheap labour markets to increase profit margins, crowdsourcing decreases job security, increases contractual employment, and requires more individuals to be 'creative' and agile in accepting odd tasks to make ends meet.

Motivational logic and/or investment of desire

Taken as a whole, the general environment of 2.0 that solicits the idle producer taps a range of motivations. For some, the task, contest, brand affiliation, ideological position, or idle expertise is motivation enough to participate; income is irrelevant. For others, however, there is a dependence on these subsistence earnings and the opportunities it provides (Ross, 2013), revealing the Janus-faced nature of participation and contribution. The tensions between these two extremes are not without political implications and have adverse effects on the labour pool, bargaining power, and working conditions (Huws, 2014, pp. 177–179). Rather than being an instance of coercive tactics or slave conditions as Fuchs suggests (2014, pp. 273–275), the enthusiasm and willingness to freely contribute seems to be a 'first world problem' of volunteering through technology, where novelty and satisfaction trump exchange-value. We willingly contribute and suffer opportunity costs in order to be part of a larger project that offers signalling incentives, social visibility, and entrepreneurial potential. The critical question still remains: Why do we continue to marginalize our labour value to receive so little in return? Phrased another way, *why did the masses desire their own repression?* (D&G, 1983).

Researchers eager to understand what keeps participants engaged evoke two common frameworks that focus precisely on rational motivations influencing individual decisions. Cognitive Evaluation Theory is a hierarchical system that takes intrinsic (enjoyment, satisfaction) and extrinsic (social norms, external pressure) motivations as the rational base of decision making. In technological environments, community, collaboration, and communication layer onto the intrinsic nature of satisfaction, through novelty, creativity, and challenges (Rotman, 2013). Social Production Function offers a more nuanced, hierarchical system, with higher 'substantive' goals (physical and social well-being) and supporting operational goals (strategic mechanisms to improve one's condition). Both are intertwined in pursuit of an improved condition, socially or economically. Within open source, signalling incentives position developers to showcase skills, create community, and provide access to new opportunities (Lerner and Tirole, 2002). Here, beliefs and an objective pursuit for improved conditions provide a way to understand the shift between primary and background goals that takes place, where acquiescence of one goal for another occurs.

Untangling rational motivations and irrational investments of desire gives a different layer to understand what motivates volunteers to participate in projects. Individual framing is the critical hinge and draws on individual proclivities, identity politics, or socio-economic circumstances. Intrinsic/extrinsic is a false dichotomy, as motivation is a hybrid of both qualities (Lakhani and Wolf, 2005; Rotman, 2013). Inducements, reward structures, and roles can be both positive and negative extrinsic motivators, depending upon orientation. Moreover, individual identification with norms, beliefs, and value systems are socially constructed. They are not definitionally fixed, but context specific on a spectrum of possible social orientations. 'Obligation' does not neatly fit into either framework (Lindenberg, 2001), making 'altruistic' explanations suspect (Lancashire, 2001).

These rational systems are also fraught with competing interests and do not capture the tension *between* the social and the individual production. Within the Web 2.0 milieu, entities actively seek to tap this production; understanding motivational rationale, whether brand affiliation, product/service evangelism, or personal interests, is key for project success. Incorporating inducements into initiatives becomes effortless and includes adding creative layers, technological efficiency, participatory role flexibility, positive feedback, contest/prize structures, or other tokens of acknowledgement (Eveleigh et al., 2013). Moreover, the desire to participate is distributed across a range of categories. Whether it is out of fun, boredom, loneliness, or a host of other rational or irrational reasons – regardless of intrinsic/extrinsic or social behavioural motivations – users actively marginalize their own labour value. The willingness to contribute defies traditional labels of 'exploitation', yet 'motivation' as rational logic is slippery. Ultimately, seeking a 'rational' explanation overlooks 'irrational' investments of desire. As Deleuze helpfully points out, 'reason' is carved out of the irrational and is defined by its relationship *to* the irrational (D&G, 2004, p. 262), rather than a pure value system.

Within the present discussion of self-marginalization, I argue that D&G's 'investment of desire', rather than 'motivation', better captures the murky terrain of forward-looking individual labour investments, where others stand to benefit from that labour. Investment of desire is both psychic and social, as desire is always assembled from the social field. Our unconscious desires and preconscious interests are in tension with the established norms and conventions that govern the social field of relations, habits, and subject identification. Desire is prior to and distinct from beliefs. While the former is a largely unmediated production, the latter is heavily mediated by the social milieu and the degree of embeddedness we have in everyday situations. Desires are channelled into existing belief systems. The more we identify with territory, ideas, practices, and customs, the more invested we are in maintaining those connections through ascription of meaning and belonging.

These individual investments are necessary, as they enable community formation and sense of belonging. This process does not exist without tension, however.

'Identification' may keep us enmeshed long after it is beneficial to be. The more we maintain those connections out of guilt, obligation, or rationalization, the more we repress desires that allow us to change over time. These previously willful relations shift into oppressive situations, or what D&G call microfascisms: investments of desire in daily practices that are adopted at the expense of present freedoms, in exchange for a belief in a better future or certainty of outcome. We subsume this oppressive condition as a rational decision with forward-looking goals: it will pay off in the future, it is the 'right' thing to do, or there is 'no choice', etc.

How this 'investment' is appealed to and captured by the social field is critical to understand how it is *transformed* into an oppressive condition: in this case, marginalizing our current labour value in exchange for potential future gain. As one example, in order to achieve a more autonomous existence, we rationalize the accompanying condition of flexible work. Some portions of our lives may be less autonomous as we become accustomed to work on demand. This exchange may seem worthwhile, as the flexibility appears more valuable than the confining workweek. This flexibility also requires taking jobs when available, creating a constant state of employability. This encroaches on work/life balance and erodes boundaries between the two. At the social level, entities appeal to desires that are activated by a set of values, beliefs, or benefits. Through inducements and rhetorical flourish, platforms like TaskRabbit, Crowdsource, Amazon's Mechanical Turk, etc., meet this desire for autonomy and creativity, *and* provide piecemeal wages and no financial security. Both employer and employee enjoy a 'just-in-time' labour contract (TEDx, 2013), yet the benefit for both is radically asymmetrical. Employers only pay for productive time, while employees must strive for hyper-efficiency. Moreover, the workforce is no longer merely the 'standing reserve' but appears to actively pursue this precarious existence.

Across the social field, the effects of this condition are agile and heterogeneously distributed. Understanding how to tap into a desire for autonomy *through* technology enables a diffuse strategy for appealing to a wider demographic, in contradistinction to appealing to held beliefs that technology *provides* autonomy. This nuanced shift of identification enables focusing on impulses that are similar across a distribution of actions.

Microfascisms 2.0

Autonomy, efficiency, and creativity are some of the common themes or impulses that drive daily practices; identity, community, and connectivity keep people enmeshed over time. Here, the desire to participate and the perceived need to stay connected creates the unique condition in which we willingly give up certain aspects of privacy or redefine work and leisure, in exchange for the seduction of what technology offers. The political imperative is precisely our willingness to participate in an environment that continues to create inordinate financial gains for a small subset of the population while volunteered labour and free production marginalizes paid labour positions. The economic differential is complex. The capitalist machine of start-ups, venture capitalists, IPOs, and acquisitions wreak havoc on local economies and residential and commercial real estate, and produce an ancillary economy that supports this production and vies to participate.

The concern here is not the Political, but the political. Microfascisms operate at the molecular or individual level as a source of 'micro-political' power. As individuals, we both give up and assert power. Active investments of desire form a supple condition – empowering or oppressive; *disengaging* from these investments can render this plane inert. Individual actions and participation infuse the social field, and *how* we choose to participate is as important as *why*. Can we disconnect or do we rationalize the need to remain connected? For D&G (1987, pp. 214–219), these moments or tensions of rationalization make up the four 'errors' of microfascism. These errors legitimate forward-looking actions, despite evidence that it may prove to be harmful for growth, change, or well-being.

The first error is to believe that 'suppleness' or deviation from norms and conventions will make things better. Desire for fun or creativity is operative here in several ways. First, filling free time with tasks framed as a fun or creative outlet provides a reprieve from otherwise unsatisfying employment conditions. Aspiring armchair scientists, scholars, or journalists contribute labour to a project out of interest, rather than selling labour as subsistence. Entities seeking labour appeal to the desire for fun, creativity, or intellectual challenge, and includes data problems, contributing labour for scientific research, writing blog entries, or coding after work for open source projects. Citizen science projects like Floracaching and Galaxy Zoo add gamification as inducement; coding tasks appeal to the creative challenge and contests inspire creative production with hope of winning a prize. Researchers acquire grant dollars and career advancement. Corporations like IBM save billions on labour costs associated with software maintenance (Tapscott and Williams, 2007), while consumer corporations acquire free market research for new products. Second, the cultural push for self-publishing or self-production as signalling incentive is another example, where individuals devote time and energy to raise visibility, writing for a high-profile site or developing an app or platform, hoping to achieve notoriety and make it an attractive acquisition or a viable public offering, like Snap Chat, Facebook, Google, and countless other tech companies. Sites require active users to assert its relevance and provide a built-in advertising audience, and thus potential value. We generate content and data in exchange for the use of free platforms, while advertising dollars line the pockets of the Technorati. Data is the new oil.

The second error is the psychological certainty and clarity of personal choices, or when we take our beliefs and investments to be our own without any external or social influences. The Web 2.0 environment is infused with a technological utopian vision and the radical democratic potential that these tools provide. This utopian spirit is particularly strong in open source. Supporters strongly identify with the belief that 'code' should be free and it offers a strong affront to capitalist business as usual (Raymond, 1999). Commitment or obligation to this community is a strong motivator to contribute free labour. Corporations have no difficulty embracing this free labour and 'deed' certain lines for public use

(Woods, 2008). *And* public maintenance. Participants sell their labour during the day, and give away their labour at night as part of community investment and identification (Lakhani and Wolf, 2005). Corporations benefit by the diversity of intelligence to keep code current, and save actual costs of employing additional talent. This enthusiasm for open source and Web 2.0 is also evident in efforts to incorporate technological efficiency into government, engendering such monikers as Government 2.0 or open government. The belief that government should be more efficient is accomplished through technology. Information apps help citizens be sensors and tap a desire for autonomy and a lightweight government. Government has the potential to reduce its payroll, while citizens become willing sensors of street conditions, utility issues, and increase participation through efficient feedback mechanisms. Open source initiatives evangelize this transition, and government seeks to reduce its own overhead by incorporating open source software.

The third error is a false security in the molecular or 'alternative' investment as a superior condition. Identifying with alternative ideas is often perceived to be more 'free' or 'pure' from conventional identification. This error builds on the second error of personal clarity, and when layered with a sense of superiority of 'alternatives', it creates a particularly insidious condition. Tech evangelism forcefully advocates for the adoption of technology, and evangelists seek to convert a larger audience with their infectious enthusiasm and energy (Lucas-Conwell, 2006). Whether it is conveying the important sociality of Facebook, the efficiency of Google search, that code should be 'free', or the flexibility of freelance, this entices users into participating through identification with the values that technology represents. Maintaining those ties is a form of community and keeps users engaged or active. This is perhaps exemplified by the running commercial, 'I'm a Mac', 'I'm a PC' in which representative personalities embody the attendant corporate spirit. To this we can add 'I'm open source' in which Android software presents a modifiable platform alternative. While Apple technology was once viewed as superior to Microsoft, embracing the more open Android platform shifts the 'alternative' status and desirability to a system that offers more flexibility. Google's business is making information accessible and search efficient; the more data it has, the more robust its ecosystem. Adding 'flexibility' and coding creativity enables Google to extend its information empire in all directions.

It is here where we can begin to see the insidious nature and influence across sectors, horizontally and vertically. These instances are not universal, but unique to each individual. Each application/opportunity taps into different aspects of our desire and transforms our present condition by appealing to efficiency, autonomy, creativity, etc., through technological innovation. We marginalize our own labour and leisure time out of desire to find a creative outlet, accumulate social capital, build skills to advance our socio-economic position, in search of community, or to possibly 'win big'. Disengaging grows increasingly difficult the more accustomed we become to the efficiency technological tools bring to our lives. Untangling motivations and desires from the reality that capitalism will indiscriminately appropriate labour production for generating surplus value is a difficult, but critical, task. It becomes even more difficult when it expands to non-profit, educational, and government institutions operating under a neoliberal economic regime. Corporations are wholly unconcerned about our 'reasons' as long as they can tap our desire to contribute, while the rhetorical flourish of 'democratic' and 'open' or 'code for good' taps ethical narratives. Each desire is simply another axiomatic, capitalistic or moralistic; adding another 'layer' of desire is effortless. As long as there is 'an app for that', free production will continue.

Conclusion

The Web 2.0 milieu automates and innovates the everyday. It taps our desires for autonomy, efficiency, and creativity, at the individual and social level. At the individual level, desires and needs to multitask, stay connected, advance social or economic positions, have greater autonomy, or inject a little creativity into our otherwise monotonous days are effortless with the diversity of Web 2.0 technologies. The agents that fuel this innovation attempt to keep apace and stay relevant with new ideas and new lines of code. At the social level, the structuring economic force metes out corporate winners and losers, requiring new business models and new philosophies of innovation; conservatism of the neoliberal environment affects government and educational agendas and installs a financial imperative of 'doing more with less'. In short, Web 2.0 technology is increasingly the answer to any problem posed, perhaps exemplified by the refrain, 'I bet there's an app for that' – whatever *that* might be.

Microfascism is a complex, robust analytic that enables an analysis of both the social forces and individual investments by focusing on the mutually constituting forces and tensions that exist *between* the two. For individuals seeking novelty, satisfaction, or community, volunteering time and expertise fills a perceived lack, provides a creative outlet, or offers new avenues of networking. The social or economic investments and signalling incentives place 'hope' on present activities for generating future gains and are coterminous with microfascisms. From the top-down social level, the desire to capture more data, access fresher ideas, subvert the corporate model, or even reassert a brand dominance creates conditions in which appealing to citizen actors is an effective strategy to maximize budgets *and* production. Users willingly marginalize their own labour value and generate free content for capitalist initiatives, research agendas, and government/non-government services. These entities strategically capture this free production through novel and creative layers, and they incentivize or appeal to individual desires to extend engagement.

What makes this environment potentially dangerous is Web 2.0's diffuse *and* hierarchical quality. It is an agile, creative movement that effortlessly links up individual desires across the social field, in which larger movements can emerge. It generates production from the ground up, but under top-down organizational schemes. The citizen actor needs a structure to insert themselves and the project 'owner' creates opportunities for participation. The Web 2.0 umbrella encompasses many efforts, interests, and persuasions and exhibits an organizing principle of individual users actively contributing *and* sustaining a variety of opportunity costs, creating a potentially oppressive situation.

Understanding the shift from empowering circumstances to enslaving conditions is both context dependent and motivationally unique, yet how it works is consistent across the social field. And it is the *scale* at which these appeals occur that should give one pause. This scale suggests the emergence of the fourth error of microfascism: believing that these two lines of production will not collapse into a *macro* formation of fascism. Every *micro*fascism is defined by what D&G call a 'micro-black hole' of subjectivity (1987, p. 224). The individual fills this hole with an investment of desire (I'm a Mac, I'm a PC . . .) and communicates with other 'black holes'. Norms and conventions affect individual practices, while immanent (individual) production communicates across the masses via processes of identification and adopting 'beliefs' or ideology, from which a larger movement emerges. These mutually constituting forces blur the pivotal transition between 'alternative' and 'compulsory' practices. The elision between empowerment and enslavement is only known after it has passed.

Celebrating this Web 2.0 environment produces fascistic tendencies, and believing that one is participating 'on *their* own terms' or to 'get ahead' through signalling incentives is a danger of the supple plane of participation. This volunteerism does not subvert the capitalist order; it only makes us more beholden to its power. *Critiquing* the participatory spirit, questioning the future of data, or challenging this utopic vision brings us to an uncomfortable moment, as adherents see this critique as a hostile affront to this radical democratic environment. 'Alternative' beliefs turn compulsory, and questions of cultural fascism thus emerge. As D&G scholar Eugene Holland (2008) posits, if a US fascist regime were possible, it would be more religious and cultural than its militaristic predecessors, which recalls both the technological evangelists and collective enthusiasm for Web 2.0 technology.

The frenzy of Web 2.0 gives a glimpse of an untenable future, one that is a dangerous acquiescence of self-worth *and* legitimate claim on our labour value. With information and data the next territory to control and the digital Technorati brokering privacy and access, larger portions of our lives are increasingly mediated by technology operating in the background. Yet the necessity of individuals selling their labour power remains, regardless of which appellation we adopt (the new proletariat, the precariat, the cybertariat) and what conceptual framing we apply (volunteered labour, free labour, exploitation, marginalization). At what point do we demand to be compensated, not because we own the idea, but because affective knowledge production does not shelter or feed us? Simply martialing a classic Marxist critique of the revolutionary potential of labour power is not enough to counter the *seduction* that technology offers. Yet what is to be done remains unclear. As Deleuze (1995) says of the Control Society, there is no need to hope or fear; we just need different (conceptual) tools, and to be agile in that adoption. What is required is a walking away from the 'hope' of a better future by confronting the present condition in which we are enmeshed. And there is no 'app' for that.

Note

1 BoardofInnovation.com exemplifies this sentiment in its online literature for prospective clients.

References

Andrejevic, M., 2013. The violence of participation. Estranged free labour. In: T. Scholz, ed., *Digital labour: the Internet as playground and factory*. New York: Routledge, pp. 149–164.

Board of Innovation, 2015. *Open innovation & crowdsourcing examples* [online]. Board of Innovation. Available at: <http://www.boardofinnovation.com/list-open-innovation-crowdsourcing-examples/> [Accessed 29 Mar. 2015].

Bowser, A., Hansen, D., He, Y., Boston, C., Reid, M., Gunnell, L. and Preece, J., 2013. Using gamification to inspire new citizen science volunteers. In: *Proceedings of the first international conference on gameful design, research, and applications*, Gamification'13 [online]. New York, NY: ACM, pp. 18–25. Available at: <http://doi.acm.org/10.1145/2583008.2583011> [Accessed 1 Apr. 2015].

Brabham, D.C., 2013. *Crowdsourcing*. Cambridge, MA; London, England: The MIT Press.

Crowdsource, 2015. *Learn how crowdsourcing works* [online]. CrowdSource. Available at: <http://www.crowdsource.com/how-it-works/> [Accessed 30 Mar. 2015].

Deleuze, G., 1995. Postscript on the societies of control. In: *Negotiations, 1972–1990*. New York: Columbia University Press, pp. 177–182.

Deleuze, G. and Guattari, F., 1983. *Anti-Oedipus: capitalism and schizophrenia*. Minneapolis: University of Minnesota Press.

Deleuze, G. and Guattari, F., 1987. *A thousand plateaus: capitalism and schizophrenia*. Minneapolis: University of Minnesota Press.

Deleuze, G. and Guattari, F., 2004. Capitalism and desire. In: Lapoujade, D., ed., *Desert islands and other texts, 1953–1974*. Los Angeles, CA; Cambridge, MA: Semiotext(e); Distributed by MIT Press, pp. 232–241.

Dijck, J. van., 2013. *The culture of connectivity: a critical history of social media*. Oxford; New York: Oxford University Press.

Dodge, M. and Kitchin, R., 2013. Crowdsourced cartography: mapping experience and knowledge. *Environment and Planning A*, 45(1), pp. 19–36.

Eggers, W.D., 2013. *Crowdsourcing social problems* [online]. Reason.com. Available at: <http://reason.com/archives/2013/12/25/crowdsourcing-social-problems> [Accessed 26 Feb. 2014].

Eveleigh, A., Jennett, C., Lynn, S. and Cox, A.L., 2013. I want to be a captain! I want to be a captain!: Gamification in the old weather citizen science project. In: *Proceedings of the first international conference on gameful design, research, and applications*, Gamification'13. [online] New York, NY: ACM, pp. 79–82. Available at: <http://doi.acm.org/10.1145/2583008.2583019> [Accessed 12 Mar. 2015].

Fuchs, C., 2014. *Digital labour and Karl Marx*. London; New York: Routledge.

Ghosh, R.A., 1998. Interviews with Linus Torvalds: What motivates software developers. *First Monday* [online] 3(2). Available at: <http://www.firstmonday.org/ojs/index.php/fm/article/view/583> [Accessed 17 Jan. 2015].

Goodchild, M.F., 2007. Citizens as sensors: the world of volunteered geography. *GeoJournal*, 69(4), pp. 211–221.

Grabar, H., 2015. *Say goodbye to the clipboard, city surveyors. Next City* [online]. Available at: <http://nextcity.org/daily/entry/city-surveys-urban-research-methods-canvas> [Accessed 29 Mar. 2015].

Holland, E.W., 2008. Schizoanalysis, nomadology, fascism. In: I. Buchanan and N. Thoburn, eds., *Deleuze and politics, Deleuze connections*. Edinburgh: Edinburgh University Press, pp. 74–97.

Howe, J., 2006. The rise of crowdsourcing. *Wired*, 14(6), p.176.

Howe, J., 2008. *Crowdsourcing: why the power of the crowd is driving the future of business*. New York: Crown Business.

Huws, U., 2014. *Labour in the global digital economy: The cybertariat comes of age*. New York: Monthly Review Press.

Lakhani, K.R. and Wolf, R.G., 2005. Why hackers do what they do: understanding motivation and effort in free/open source software projects. In: J. Feller, ed., *Perspectives on free and open source software*. Cambridge, MA: MIT Press, pp. 3–21.

Lancashire, D., 2001. Code, culture and cash: The fading altruism of open source development. *First Monday* [online] 6(12). Available at: <http://firstmonday.org/ojs/index.php/fm/article/view/904> [Accessed 5 Mar. 2014].

Lazzarato, M., 1996. Immaterial labour. In: P. Virno and M. Hardt, eds., *Radical thought in Italy: a potential politics*. Minneapolis, MN: University of Minnesota Press, pp. 133–147.

Lazzarato, M., 2012. *The making of the indebted man: an essay on the neoliberal condition*. Los Angeles, CA: Semiotext(e).

Lerner, J. and Tirole, J., 2002. Some simple economics of open source. *The Journal of Industrial Economics*, 50(2), pp. 197–234.

Lindenberg, S., 2001. Intrinsic motivation in a new light. *Kyklos*, 54(2/3), pp. 317–342.

Lucas-Conwell, F., 2006. Technology evangelists: a leadership survey. SDForum Conference on Technology Leadership and Evangelism in the Participation Age Available at: <http://www.growthresourcesinc.com/pub/res/pdf/TechEvangelist.pdf> [Accessed 20 Feb. 2016].

Pasquinelli, M., 2009. Google's pagerank algorithm: a diagram of cognitive capitalism and the rentier of the common intellect. In: K. Becker and F. Stalder, eds., *Deep search: the politics of search beyond Google*. Innsbruck; Piscataway, NJ: Studien Verlag; Distributed by Transaction Publishers, pp. 152–162.

Raymond, E., 1999. The cathedral and the bazaar. *Knowledge, Technology & Policy*, 12(3), pp. 23–49.

Ross, A., 2013. Looking for the lost paycheck. In: T. Scholz, ed., *Digital labour: the Internet as playground and factory*. New York: Routledge, pp. 13–32.

Rotman, D., 2013. *Collabourative science across the globe: the influence of motivation and culture on volunteers in the United States, India, and Costa Rica* [online]. Digital Repository at the University of Maryland. Available at: <http://hdl.handle.net/1903/14163> [Accessed 29 Mar. 2015].

Standing, G., 2012. The precariat: From denizens to citizens? *Polity*, 44(4), pp. 588–608.

Tapscott, D. and Williams, A.D., 2007. Innovation in the age of mass collabouration. *BusinessWeek: Innovation_and_Design* [online]. 1 Feb. Available at: <http://www.businessweek.com/stories/2007–02–01/innovation-in-the-age-of-mass-collabourationbusinessweek-business-news-stock-market-and-financial-advice> [Accessed 25 Mar. 2014].

TEDx StLouis Women, 2013. *Connecting the world's workforce: Stephanie Leffler.* Available at: <https://www.youtube.com/watch?v=VBhxy8z-0Hw&feature=youtube_gdata_player> [Accessed 31 Mar. 2015].

Terranova, T., 2000. Free labour producing culture for the digital economy. *Social Text*, 18(2), pp. 33–58.

Weber, S., 2004. *The success of open source*. Cambridge, MA: Harvard University Press.

Woods, D., 2008. The commercial bear hug of open source [online]. *Forbes*. Available at: <http://www.forbes.com/2008/08/17/cio-open-source-tech-cio-cx_dw_0818open.html> [Accessed 1 Mar. 2014].

Openings

14 Between the Fool and the World

Towards a (re)contextualization of assemblage thinking in the contemporary university

Darren J. Patrick

> Worry about the university. This is the injunction today in the United States, one with a long history. Call for its restoration like Harold Bloom or Stanley Fish or Gerald Graff. Call for its reform like Derek Bok or Bill Readings or Cary Nelson. Call out to it as it calls to you. But for the subversive intellectual, all of this goes on upstairs, in polite company, among the rational men. After all, the subversive intellectual came under false pretenses, with bad documents, out of love. Her labor is as necessary as it is unwelcome. The university needs what she bears but cannot bear what she brings. And on top of all that, she disappears. She disappears into the underground, the downlow lowdown maroon community of the university, into the undercommons of enlightenment, where the work gets done, where the work gets subverted, where the revolution is still black, still strong.
>
> Fred Moten and Stefano Harney, 'The University and the Undercommons' in *The Undercommons: Fugitive Planning & Black Study* (Minor Compositions/Autonomedia, 2013)

Preamble: reading differently

- *I'm trying to make sense out of this nonsense.*
- *We need a guide; or two.*

Some things about me: I'm a queer (faggy) white guy from the working class; I'm the first one in my immediate family to earn an undergraduate degree and to pursue post-graduate studies. I feel foolish reading Deleuze and Guattari. Masking that foolishness is part of getting by in the academic hierarchy. No doubt it is cool – if no longer vogue – to speak in their unique language. I get it? I get it. I get it! I'm over it. If I'm honest, I don't want to fight to be a part of the boy's club that so often forms around their work in the Anglo-American academy. But there are other paths, other ways into their world. I pick a card and try to read differently (Figure 14.1). The Fool.

The Fool is the only unnumbered card in the Tarot de Marseille. In other decks it is numbered zero; the sign-symbol is a metonym for the orphic egg. The Fool is placeless and numberless. It is the card of beginnings and endings, alternately

Figure 14.1 The Fool and the World.
Source: Assemblage by Darren J. Patrick.

counted as the highest or the lowest of the Trumps. The Fool is fully between – 'dark precursor' (Deleuze, 1994, p. 119) – outside within. The Fool ascends and descends the order of the Tarot, staging encounters that mark out the worldly and spiritual dilemmas of the other Trumps. The Fool is that character who, being numbered and nothing at the same time, seemingly coordinates the whole system of encounters and conjunctions that help the willing participant work with the Tarot (Semetsky, 2001).[1] The Fool takes the initial steps on a relational drama that unfolds across the Major Arcana, or the first twenty-one cards of the Tarot.

On the one hand, the Fool is a figure of great openness to the multiplicity of encounters which will ensue on the journey and, on the other, the Fool is a figure of great stupidity, even madness, unaware of its direction, nomadic, dangerously unhinged, in need of a world. This Fool is among the group, but detached; this outsider within embodies the tense relationship between a point of origin/departure and utter detachment from any specificity or belonging. The Fool is a Simmelian stranger, one who marks a certain margin.

Reaching this margin, I've got to ask: Are Deleuze (and Guattari), as Peter Hallward (2006) suggests, calling us out of this world, away from the everyday struggles of concrete politics? When we work with them, create with them, are we *necessarily* working on making this world more just and survivable for everybody? Do we relinquish something – injure something – when we take up that call to '[w]rite, form a rhizome . . .' while forgetting (or failing to grasp) the second half of that exhortation: '[I]ncrease your territory by deterritorialization, extend the line of flight to the point where it becomes an abstract machine covering the entire plane of consistency' (Deleuze and Guattari, 1987, p. 10)?

Are these concepts still bricks for breaking windows? Or, have they become bricks for building a yet another version of institutionally sanctioned criticality? Certainly, it depends on the arm that throws, the specific gravity of the situation, who cowers inside. Beyond the mythology – to be clear, I think we are never wholly beyond the mythology – it is easy to be seduced by the complexity of Deleuze and Guattari's special form of foolishness; how can they walk in so many worlds at once? Be open to so many things? It takes a certain kind of courage to write about things you've never experienced. I think they squinted and saw the outside from within; their world was not infinite, as it is so easy to assume. Not all texts were on their table.

Surely, their project cannot be reduced to this. It is not-one. I'd like to think it is an invitation to reconsider those institutions and structures which (al)most (in)visibly bind us to particular patterns of working, thinking, and being. These patterns are rooted in the fear of really doing what we claim to want to do when we do critical theory; fear of being foolish in the face of (minor) differences; fear which, as Sara Ahmed (2012) has it, leads some to think we have 'gotten over' categories like race and gender through the use of terms such as 'assemblage' (p. 180). How do these categories live in me? In you? How do they become familiar? What happens when a call to defamiliarize ourselves becomes *too* familiar? What happens to the big and banal violences, to the grand and little fascisms, to the worlds we leave behind as we round the corners of our foolish journey?

- *We should be excited and suspicious of the Fool's journey.*
- *We can follow, but a few steps behind, shaken loose in the wake of their walk.*

Pick again. The World. The Fool's journey through the Major Arcana culminates in *le Monde*, the World, sometimes called the Universe. This is a card of accomplishment and completion. On it, we find an androgynous human figure, dancing at the centre of a yonic wreath. In the Tarot de Marseille, the wreath is surrounded by four elemental figures, called the tetramorph: the cherub/human figure (Aquarius/air), the eagle (Scorpio/water), the lion (Leo/fire), and the bull (Taurus/earth). Assembled together, the figures are joined in celebration of the historical accomplishment of the Fool's journey-ordeal. This is a card of triumph in the formation of a world. The World is both an abstract and an earthly card. It is practical, a card which crowns cyclicality, renewal, and unity. It is at once emancipatory and inaugural, suggesting both the notion of completion and the inevitability of return. The World is a card of rhythms and wholeness. It suggests a freedom guided by the teacherly values of responsibility, discipline, and contemplation. The World, in a word, is totality. Totality in every minor instance.

- *Why are we talking about the margin?*
- *Whose margin is it anyway?*
- *I guess that depends on who is here to claim it.*

Introduction: context of contexts

This chapter calls for (re)contextualiztion of assemblage thinking in two specific ways: first, by plotting some points on a map of the political and intellectual milieu from which the oft-cited concept of *agencement* could be elaborated by Gilles Deleuze and Félix Guattari; and, second, by articulating some (institutional) political implications of the rise of assemblage thinking as a mode of addressing marginality. Why? Quite simply because what was arguably a radical and iconoclastic concept in post-war French philosophical, psychoanalytic, and political-critical circles risks becoming, in its various appropriations, translations, and transatlantic travels, a somewhat fetishized mode of criticality among university-based scholars.[2]

In the first approach, I am largely concerned with necessarily selective historical and biographical details that help to situate Deleuze and Guattari both in relation to each other and in relation to the (non-/counter-/para-)institutional contexts that enabled their writing, thinking, and action *as a marginal philosophical project* that emerged in the wake of the mass political upheaval of May 1968. My approach to the second context more fully elaborates the importance and complexity both of translating and appropriating the philosophical concept of *agencement* and of considering the impact of *other* 1968s with respect to experiences of and claims to thinking 'marginality' as a radical political project in the Anglo-American academy today. My claim is that thinking *agencement* with/through feminist, anti-racist, and queer critiques of marginality might more fully address the stakes, promises, and perils of the institutionalization of minority and/or marginal projects, people, and positions in the university in the present conjuncture.

Between these two contexts, I am less interested in the world of concepts that emerges from Deleuze and Guattari's work than I am in the *mondialization* of their thought,[3] that is, the movement of their naïve concepts into the world. My hope is that my intervention will accomplish two things. First, that it will temper the nascent deification of Deleuze and Guattari as unmarked markers of criticality. This tendency is, of course, reproduced in renderings of *agencement* that unfold in a closed/canonical philosophical citational economy, one which valorizes Europe's self-enlightening discovery of difference. I favour a more biographically contingent and praxis-based reading of their work. And, second, that it will subvert a general tendency of unmarked white, male, heterosexual, dominant – in short, 'major' (Deleuze and Guattari, 1987, pp. 104–105) – critical theory to treat feminist, anti-racist, or queer – in short, 'diverse' or 'minority' – interventions[4] as 'value-added' moments for ongoing reproductions and extensions of dominant interpretations and appropriations of European philosophy, critical or not (Alexander, 2005, pp. 138–139). Here, I am struggling against these add-on politicizations and phagocytic incorporations.

Arrangement, *agencement*, assemblage

As François Dosse (2011) tells it in his landmark double biography (on which this section draws heavily), Gilles Deleuze and Félix Guattari's joint philosophical project emerged out of an unlikely friendship between an heretical philosopher and a militant psychoanalyst. The pair's collaborative development of a dizzying array of concepts cannot be separated from the very arrangement that guided their friendship and collaboration: 'Their writing is organized so as to put into place a collective *arrangement* of utterance. . . . It is essential to understand the notion of *arrangement* to appreciate how unusual the situation was' (p. 10, emphasis in original). *Agencement* – loosely, even poorly, translated as assemblage or arrangement – is a thematic that insists on the 'temporary . . . encounter of a number of elements . . . that create something resembling a context that bears on, for example, the perception of an object' (p. 527 fn43). Their arrangement relied on several well-defined rules set for Guattari by the pedagogically inclined Deleuze: a strict writing schedule, daily drafts, and a weekly meeting. The rules were designed to help Guattari overcome his writer's block and to facilitate the mutual opening of each other's particular talents, skills, and experiences both to each other and to their world (p. 7).

Alongside Deleuze's studies of individual philosophers,[5] Guattari's insistence on the importance of empirical studies, militant activism, and institutional analysis significantly informed the character of their collaborative works. In 1965, Guattari helped to launch the Federation of Institutional Study Groups and Research (FGERI) and the affiliated Centre for Institutional Study, Research, and Training (CERFI), which founded the journal *Recherches*, whose inaugural issue was published in 1966. Together, these groups put Guattari's 'labordian' (p. 62) concept of transversality into practice by linking academic disciplines not only to each other but also to non-academic professions (p. 78).

Transversality is an extension-actualization of the practice of institutional psychotherapy, of which '[o]ne of [the] major tenets was that mentally ill patients could only be treated in an institution that reflected on how it operated itself' (p. 60). The other major tenet of institutional psychotherapy was that 'treatment' must unfold through 'the introduction of new arrangements and social connections' (p. 60). In this context, Guattari sought to support the enablement of self-determined group-subjects at La Borde, as opposed to fully structurally determined 'subjected groups' (pp. 60–61). Although various structurally legible groups converged at the clinic – militant students of Guattari's and Lacan's, patients, medical staff, and non-medical staff – the institution itself relied on many overlapping and heterogenous small groups both to manage and to transform everyday life at the clinic. Rather than adhering strictly to naturalized categories, the groups were aimed specifically at creating new subjectivities for *all* people involved in the clinic's work.

Transversality, then, named both the goal and the method of this process, which was to, 'unsettle binary structural oppositions . . . mak[ing] it possible to start the analytical process of getting out of oneself and the necessary shift in confronting the group' (p. 62). The combination of institutional psychotherapy and transversality shifts attention away from the *knowledge of* a pathologized subject's 'condition' or 'illness', instead emphasizing the concrete practices that might actually transform the conditions under which the individual is reduced to a marginalizable pathology, especially in an institutional setting. As the second section will elaborate, the tension between self-determination and institutional location can, with the combined, but differential, efforts of systemic beneficiaries and marginalized groups, transform oppressive and hierarchical institutional relations by way of transversal group-formation; today, these might be called alliances. Resources, risks, and responsibilities must be shared in an everyday and concrete way such that (a) roles are not predefined according to disciplinary specialty or social location and (b) the real and accumulated effects of training, privilege, and power are not artificially flattened. Collectivization, in other words, is necessary to change institutional arrangements in which the ostensibly 'empowering' dimensions of (quasi-)fascistic tendencies are individualized on the 'majority' body while the effects of such (quasi-)fascisms are wrought on 'minority' bodies.[6]

As will become clear in the next section, I am keenly emphasizing Guattari's work because it speaks to the very necessity *not* to reproduce the concept of *agencement* apart from a rigorous understanding of the conditions under which that concept is promulgated, made portable as a methodology, and assumed to be capable of teaching us something 'about the world' beyond heavily institutionalized relationships. Deleuze, of course, already translates Guattari's clinically-based concept-practices into the university by rendering them, to a certain extent, through his own philosophical project.[7] Here we cannot forget that the arrangement which guided their actual collaborative work would never have formed were it not for the 'founding rupture' of May 1968 in France (Dosse, 2011, p. 178). May 1968 is the link by which Guattari and Deleuze articulate their *agencement*; the link through which the institutionalized practice of philosophy in the university meets praxis in the clinic; the link by which the dedicated teacher meets the frenetic practitioner; the link through which willful heretic meets the scattered, anxious militant; the link between two fools and many worlds.

As Kristin Ross (2002) has meticulously documented, the political claims and implications of May 1968 *in France* have too frequently been co-opted, depoliticized, and presented outside of the context both of the political work that preceded them and of their consequences. Nonetheless, the events can be summed up thusly:

> May was a crisis in functionalism. The movement took the form of political experiments in declassification, in disrupting the natural 'givenness' of

> places; it consisted of displacements that took students outside of the university, meetings that brought farmers and workers together, or students to the countryside – trajectories outside of the Latin Quarter, to workers' housing and popular neighborhoods, a new kind of mass organizing (against the Algerian War in the early 1960s, and later against the Vietnam War) that involved physical dislocation. And in that physical dislocation lay a dislocation in the very idea of politics – moving it out of its proper place, which was for the left at that time the Communist Party. (p. 25)

Indeed, the general atmosphere of dislocation, the formation of new group alliances, and of the betrayal of appropriate (institutional and political) positions did not originate solely in the mythically celebrated moment of May 1968. This work was also articulated through de/anti-colonial and workerist politics (Ross, 2002, p. 9), extensions of which were ultimately expressed in the student movement and general strike. That the subversive political 'difference' championed in dominant remembering of May 1968 in France becomes part of the conceptual-philosophical difference that undergirds concepts like *agencement* is a key justification for tracing some of the subterranean historico-political links between the French and Anglo-American academic contexts.[8]

Given the anti-institutional nature of the events of May 1968 in France, Guattari's militancy in mobilizing networks from La Borde – as well as those of FGERI and CERFI – eventually led his colleague, Jean Oury, to ostracize him from the clinic (Dosse, 2011, pp. 175–176). Meanwhile, Deleuze, while in solidarity with the events, was focused on finishing his doctoral thesis, *Difference and Repetition* (1994). This text marks a shift from his studies of individual philosophers to his overall project of overturning Platonism in favour of a differential philosophy of the multiple. Later, through his collaboration with Guattari, core concepts from both *Difference and Repetition* and *The Logic of Sense* (Deleuze, 1990) were fused in a mutual effort to move past dominant theories of Lacanian linguistic structuralism in which unconscious functions are referred back to the Oedipal complex and interpreted through familial triangulation.[9] In light of the political context, their effort was not given. It instead resulted, in part, from Guattari's drift away from Lacan, who, at a certain point, favoured Guattari as a 'successor' and interlocutor in the manner of patriarchal patronage of the French academy (Dosse, 2011, pp. 223–224), and, also in part, from his fortuitous collaborative arrangement with Deleuze, whose ideas gave philosophical ballast to Guattari's profoundly rich political and militant experience. In the end, Guattari 'wanted to draw lessons from May '68 and reenergize the structures that had been shaken up by those events' (p. 225).

Marginality, difference, diversity

With the deliberately selective and inevitably partial biographical and historical sketch of Deleuze and Guattari's reading-writing arrangement now on the table, this section will identify several axes of intersection and points productive betrayal that arise if we replace *agencement* (assemblage/arrangement) in the context of putatively 'marginal' struggles waged by feminist, anti-racist, and queer thinkers in the contemporary Anglo-American university.

By way of conjunctural connection, I begin again with upheavals of 1968 which were, of course, not confined to France. As Roderick Ferguson (2012) tells it in his astounding study, demands for the recognition of the productive role of difference – in experience, position, history, identity, and desire – were central to *global* social movements of 1968. Indeed, we must betray Eurocentric renderings of that year and refuse to interpret the 'recognition of difference' as an outgrowth of progressive European enlightenment. Ferguson writes:

> While difference in the French context often meant the suppression of the worker and the colonial militant as revolutionary figures, difference in the U.S. setting meant calling for the presence of those subjects deemed to be outside the American academy and its systems of representation. Difference during the sixties in the United States would become the moment in which minoritized subjects of all types would brim with versions of that question that Angela Davis asked herself upon arriving on the UCSD campus: "Where are my people?" (p. 49)

Indeed, emphasizing the non-identical/differential meanings of the ruptures of 1968 – which Davis's words mark as an absence – is Ferguson's starting point for an analysis of what exactly is at stake in contradictory institutional affirmations of minority difference *as* marginality. The very incorporation of minoritized bodies into the university speaks of the institutionalized will-to-knowledge of minority difference and marginality as a strategy of so-called affirmative action, a strategy by which 'the institution will now reward those who have overcome their adverse circumstances with the resources of individual character' (Butler, 1998 quoted in Ferguson, 2012, p. 199). Simultaneously cultivating a lack of faith in the university's liberating potential and recognizing it as a site of struggle learns from the troubling and contradictory institutionalizations of anti-racist, de-/anti-colonial, and feminist social movements in what Ferguson calls the 'drama of affirmation'.

In the chapter that bears this title, Ferguson draws *inter alia* on Gayatri Spivak's work and Jhumpa Lahiri's novel *The Namesake* to identify how marginality was institutionally constituted as a quasi-medicalized condition to be managed and investigated as an individual pathology in the context of the liberal (and multicultural) university. Ferguson concludes, 'Political emancipation would become the spirit of hegemonic institutionalization – in which modes of minority difference would be selectively incorporated, but not for the good of social redistribution' (p. 174). Indeed, the formation of a group-subject or collective enunciation, to use Deleuze and Guattari's lexicon, was far from immune to bureaucratic depoliticization and cooptation.

And yet, the will-to-institutionalization is, of course, never fully constitutive of the worlds which it seeks to manage and to make legible; it is not a totalizing gesture. The excesses of this gesture are revealed if we look at the university's emergent administrative and bureaucratic relationship to queer sexualities, whose institutionalization represents the latest scene in the drama of affirmation. Ferguson points out, 'To conceptualize sexuality as an institutional and administrative operation means that sexuality partakes in encounters that we have yet to theorize' (p. 223). It follows that subverting the seductions of the will-to-institutionalization is, in part, the properly political work of 'assembling life at the margins'. For Ferguson, this is a matter of *rereading*, *rewriting*, and *redesigning* (without rereproducing) the very contexts in which apparently specific forms of minority difference appear to mark out zones/objects of critical inquiry as both legible and legitimate (p. 230). Rather than simply increasing critical territories for psychic and epistemic colonization, the work of assembling life at the margins must encompass a practice of attending to 'the little things that we can deploy in order to imagine critical forms of community, forms in which minoritized subjects become the agents rather than the silent objects of knowledge formations and institutional practices' (p. 232). The point, then, is precisely *not* to put methodologies or concept-practices to work in the service of legitimizing 'difference' so that it may, yet again, be consumed, but to recognize that encounters which elude capture by institutionally sanctioned forms of enunciation are those which continue to hold potential for upsetting the smooth functioning of the affirmative machine that is the Anglo-American academy.

Audre Lorde's (1993) perennially powerful 'The Master's Tools Will Never Dismantle the Master's House' reminds us that the work of bearing witness to the everyday operations of power and oppression is a necessary part of holding the line against the seductions of institutional affirmation. Lorde spoke the words of 'Master's Tools' as an intervention at the Second Sex Conference, held in 1979 at New York University, for which she had been invited to comment on papers by what turned out to be the only two women of colour at the conference. Both panelists had been found at the last minute; they had been included only to shoulder the entire burden of speaking of 'difference' in feminism. Lorde writes/speaks:

> Advocating the mere tolerance of difference between women is the grossest reformism. It is a total denial of the creative function of difference in our lives. Difference must not merely be tolerated, but seen as a fund of necessary polarities between which our creativity can spark like a dialectic. . . . Within the interdependence of mutual (nondominant) differences lies the security which enables us to descend into the chaos of knowledge and return with true visions of our future, along with the concomitant power to effect those changes which can bring that future into being. *Difference is that raw and powerful connection from which our personal power is forged.* (pp. 111–112, emphasis added)

And yet, we find that institutionalized renderings of 'marginality' are yet another form of hailing just such 'differences' only in order to maintain a 'safe', and often policing, distance. The point is not, however, to discard such renderings so much as to attend to their specific operations. Revisiting Lorde's intervention, Micah White (2009) makes a deft connection here: '[I]f we learn anything from Gilles

Deleuze and Félix Guattari's exceptional *A Thousand Plateaus: Capitalism and Schizophrenia* it should be that the dominant powers appropriated from us first' (para. 6). Let's not, then, swap one 'master' for another or stumble foolishly into a world we may know too little about as if the problem is, precisely, our lack of knowledge. Knowing ourselves and our complicities with the institutions which may house us is a prerequisite for enacting meaningful and concrete change.

Two decades on and just a few blocks from where Lorde's words again called into being the collective arrangement that would seek to transform 'marginalized difference' into justice and *truthfully* radical change, a group of students at the highly regarded institution for European Studies, the New School for Social Research, formed the Mobilization for Real Diversity, Democracy, and Economic Justice. As M. Jacqui Alexander (2005) documents, the movement occurred between 1996 and 1998 and sought 'to transform a wide range of inequitable institutional practices in hiring, decision making, curricular and intellectual projects, and the unequal appropriation of labor, all of which produced a culture of domination that stood in the way of justice' (p. 118). In short, the Mobilization actively critiqued and attempted to dismantle both the epistemological basis for and the practical outcomes of tokenizing and exploitive 'diversity' policies, practices, and rhetorics. The Mobilization's crystallizes a notion of *agencement* in institutionally concrete terms: The centering of diversity/marginality *as critical theory* means nothing without an end to the everyday practices of white supremacy, sexism, homophobia, and exploitation in the university. The bitter ironies that obtain when this fails to happen are only stiffened when an institution thrives on a 'radical' reputation.

Tracing a line from 1968 to Lorde to the Mobilization speaks of the real, practical challenges to realizing the collective and enunciative conditions for social justice in apparently radical spaces of institutionalized racism, sexism, homo/transphobia, and classism. The New School for Social Research (NSSR), after all, had built its corporate 'brand' on its historical role as the University in Exile, 'which in 1934 was comprised almost entirely of German exiles fleeing fascism' (p. 119). As the administration of NSSR sought to capitalize on this reputation in an increasingly liberal and globalized market economy, the Mobilization worked to reveal and to redress the commodification and management of 'difference' through 'diversity' policy. Their work revealed that the particular, although non-unique, diversity policies of NSSR indexed and audited 'minority' and 'marginal' bodies absent any 'commensurate discursive structures [to] spectacularize or rank whiteness'. Operating through the left-liberal logics of inclusion/affirmation, the policies consistently failed to offer anything other than revolving doors and exploitative conditions for faculty of colour, women, LGBTQ faculty and staff, and food service and security workers.

The institutional critiques of the Mobilization emerged, in part, through a proposal to retain then-visiting scholar M. Jacqui Alexander as a permanent faculty member in Gender Studies and Feminist Theory. As Alexander writes, 'It is through individuals that structures come to light. But structures themselves are never reducible to individuals' (p. 125). Indeed, the transversal connections and institutional critique opened by the Mobilization created a space for resisting the emergent counter-Mobilization's attempt to pathologize/individualize 'the problem' of difference – and, therefore 'the solution' of diversity/affirmative action – onto a single body or collection of enumerated bodies. Nevertheless, a proposal put forward by the Mobilization – which demanded a systematic review of the retention of faculty of colour, including Prof. Alexander – recognized that such moves would enable broader transformative discourses on the reproduction of whiteness and the 'system of segregation that absents white men in power, even while they rule' (p. 140). The dismantling of such an arrangement is necessarily an epistemic, worldly, spiritual, and practical labour that not only 'affirms' or 'appreciates' '*non*-European' projects, but also *actually* revives a legacy of anti-fascist theorizing by 'present[ing] an opportunity to link the rise of fascism to colonialism and racism' (p. 149, emphasis in original). The Mobilization, in other words, embraced a concept-practice of difference through which there could be 'a theoretical engagement of particular knowledges, which were linked to the intellectual projects of scholars of colour, not [just the] ritual inclusion of either bodies or texts' (p. 166).

Conclusion: inclusion is not the end

Agencement seeks to enable the elaboration of difference en route to contextual understandings of the institutionalizations of power and the potential for unmasking and unsettling the oppression of marginal bodies, groups, and practices. It has been my argument that, if *agencement* is to be a critical, even, dare I say, liberating, concept-practice, then we must skeptically attend to the particularities of the very institutions that profess commitments to diversity and difference, including through the elaboration of ideas of assemblage and through the gesture of assembling life at the margins. As Sara Ahmed (2012) reveals, although many universities are increasingly managing to 'make room' for 'diverse' bodies through commitments to inclusion, such principled policies often, and necessarily, fail *precisely* on their own terms.

This drama unfolds alongside ever more ambitious theoretical projects that often and continuously fail to take seriously their own complicities with relations of power and oppression that structure the university itself as a site of critical enablement. As Ahmed puts it:

> The hope invested in new terms (mobilities, becomings, assemblages, capacities) can thus be considered a way of overing, as if these terms allow us to get over the categories [like gender and race] themselves. In turn, academic work that works on questions of gender or race or that works with existing social categories . . . becomes associated with stasis. (p. 180)

Ahmed is not, as she makes clear, 'saying that we need to dismiss these new theoretical vocabularies' (p. 181). Instead, she is asking that we struggle to avoid reinscribing 'the *narrowing* of the descriptive or analytical potential of old terms' such as gender, race, or sexuality (p. 181, emphasis in original). Instead, we might think our reading and writing practices in ways not fully reducible to institutional logics. Such alternative practices hold latent potential for the productive betrayal of the orthodox repetition of heterodox idea(l)s.

Here, at the end of a chapter at the end of a book on rethinking life at the margins, I suggest that life – our own lives and the lives of 'others' – has as much to teach us about difference as do theories rooted in attempts to think the operation of difference-in-itself. Theories, like Deleuze and Guattari's (1987), that used many of the best critical resources available to them to call us towards a serious engagement with 'minor languages' which, for them, 'do not exist in themselves' but 'exist only in relation to a major language and are also investments of that language *for the purpose of making it minor*' (p. 105, emphasis added). Indeed, *agencement* might stand for our encounter with the world, an encounter in which we should face our foolishness and recognize that '[i]t is not the time to be over it, if it is not over. It is not even the time to get over it' (Ahmed, 2012, p. 181).

With this in mind, I ask: What is assemblage for? What might *agencement* be able to do and to mean if it were not a substitute or metacategory standing in for historically specific claims to *being human* despite divisive forces of antagonistic differentiation? The university must cease to be a space which devours minoritized languages and lives to strengthen or elevate 'major' theory. It must, instead, thrive by humbly marking the minor details of the life of theory itself. *Agencement* may or may not help us think and act our way to making the university – and the lives we lead inside and in spite of it – a more just, collectively empowering, and non-hierarchical point of departure.

Notes

1 If we follow Inna Semetsky's (2010) Deleuzian interpretation: 'The transversal communication created by Tarot enables one to potentially cross the threshold of one's old habitual universe of thinking and acting by forming a "self–reflexive echo chamber, a feedback loop" embedded in the semiotics of experience' (pp. 91–92).

2 Indeed, there is no shortage of university-based thinkers who don Deleuze and Guattari's (1994) conceptual personae for all manner of projects, creating a veritable world of uses and applications. One need only think of the University of Edinburgh's 'Deleuze Connections' series, which has produced more than 25 titles following the formula 'Deleuze and . . .' (see, for example, Nigianni and Storr, 2009; Bonta and Protevi, 2004). In addition, Deleuze and Guattari have been invoked through queer theory (Ruffalo, 2009), theorizations of urban space and politics (Farías and Bender, 2010), and exploring the virtualities of scale through global cities (Ong and Collier, 2005). My point is neither to disparage nor to discredit such elaborations – indeed, I do not systematically engage this literature at all here – but to pose a rather simple question in light of this mini-industry: why these ideas here and now? What potentials do they hold for creating or supporting concrete change?

3 In Henri Lefebvre's Kotsas Axelos–inspired sense, *mondialization* is 'the process of becoming worldwide, the seizing and grasping of the world as a whole, comprehending it as a totality, as an event in thought' (Elden, 2008, p. 80). As Elden writes, we might consider that '[t]otality is an aspiration rather than an intellectual step in the process of thought' (p. 85).

4 For me, reading and doing feminist, queer, and anti-racist work has helped to emphasize the importance and complexities of lived struggles of identity and differences to all theoretical projects. Canonization, then, refers to the process by which a worldly project gets to claim access to the World, and is thereby allowed to stumble foolishly into all manner of other worlds, putting them to use in ways that often subvert (claims to) an emancipatory politics. Such a risk is frequently reproduced without regard to the reality that such worlds are inhabited, made, alive, in a word, also human.

5 Deleuze's early work was a formal, and somewhat begrudgingly undertaken, training exercise in upending a version of the European philosophical canon by engaging in the work of 'recreat[ing] the singularity of his predecessors' while 'stashing bombs beneath the pedestals of each master thinker'(Dosse, 2011, p. 108).

6 I am deeply indebted to members of my immediate scholarly community at York University, particularly those of the Accessibility, Community and Equity @ York (ACE) group and members of the Equity Committee in the Faculty of Environmental Studies, for modeling, teaching, and sharing these principles during still ongoing struggles to resist white supremacy, colonialism, ableism, and homo and transphobia in our Faculty and at the University.

7 This, in part, is why it is so easy to 'forget' Guattari, as often happened when their first collaborative work was released and reviewed (Dosse, 2011, p. 208).

8 Nonetheless, there are also textual travels which we should not forget. Dosse (2011) credits Sylvère Lotringer, who launched Semiotext(e) in 1974, for helping to diffuse Deleuze and Guattari's work in New York City and the US (p. 466). Indeed, Lotringer's friendship with Guattari stands as something of a key to the dominance (in the US, at least) of a 'molecular' interpretation of both May 1968 in France and the autonomous movements that swept Italy in 1977. Canada-based academics also had a key role to play in the translation and diffusion of Deleuze's work and, eventually, Deleuze and Guattari's work: Constantin Boundas translated four of Deleuze's solo works (p. 471). Along with Paul Patton, Brian Massumi, and others, Boundas brought Deleuze and Guattari to a wider Anglo-American audience in the early 1990s. Considering the dominance of analytical approaches in philosophy departments in the US academy, Deleuze and Guattari's ideas took flight elsewhere: in cultural studies, film studies, and, through Eleanor Kaufman at Duke University, theological studies (pp. 474–475). In the latest wave of Deleuze and Guattari enthusiasm, Brian Massumi's contributions, along with those of Manuel DeLanda, are central (p. 477). Lastly, and perhaps most importantly for the relationship to radical politics in/of the Anglo-American academy, it is the Spinozist post-Marxist work of Toni Negri and Michael Hardt that has given a particular spin to Deleuze and Guattari's reception in terms of radical political potential (p. 477). Interestingly, although Dosse is at pains to draw links among Deleuze and Guattari's work, feminism (Donna Haraway, Elizabeth Grosz), and post-colonial studies (Gayatri Spivak), he does so only monodirectionally, while avoiding internal differentiations in both fields and both fields' close relationship to campus and university activism in the US and elsewhere from the 1960s onward (pp. 477–478). In general, Dosse's approach to theorizing the 'export' of Deleuze and Guattari does not question the very manner in which the assemblage of Deleuze–Guattari–May–1968–in–France impacts their standing and interpretation in the Anglo-American context, where 1968 had its own meanings, political aspirations, and concrete, if contingent, outcomes.

9 Here, we would do well to remember Lacan's reaction to 1968: 'Lacan, for one, went so far as to comment about the demonstrators that they were lost souls in search of a father or "aspiring toward a master"' (Ross, 2002, p. 191).

References

Ahmed, S., 2012. *On being included: racism and diversity in institutional life*. Durham, NC: Duke University Press.

Alexander, M.J., 2005. *Pedagogies of crossing: meditations on feminism, sexual politics, memory, and the sacred*. Durham, NC: Duke University Press.

Bonta, M. and Protevi, J. eds., 2004. *Deleuze and geophilosophy: a guide and glossary*. Edinburgh: Edinburgh University Press.

Deleuze, G., 1990. *The logic of sense*. Translated from French by Mark Lester with Charles Stivale. London: The Athlone Press.

Deleuze, G., 1994. *Difference and repetition*. Translated from French by Paul Patton. New York City: Columbia University Press.

Deleuze, G. and Guattari, F., 1987. *A thousand plateaus: capitalism and schizophrenia*. Translated from French by Brian Massumi. Minneapolis, MN: University of Minnesota Press.

Deleuze, G. and Guattari, F., 1994. *What is philosophy?* Translated from French by Hugh Tomlinson and Graham Burchell. New York City, NY: Columbia University Press.

Dosse, F., 2011. *Gilles Deleuze and Félix Guattari: intersecting lives*. Translated from French by Deborah Glassman. New York City, NY: Columbia University Press.

Elden, S., 2008. Mondialisation before globalization: Lefebvre and Axelos. In: K. Goonewardena, S. Kipfer, R. Milgrom, and C. Schmid, eds., *Space, difference, everyday life: reading Henri Lefebvre*. New York City, NY: Routledge, pp. 80–93.

Farías, I. and Bender, T. eds., 2010. *Urban assemblages: how actor-network theory changes urban studies*. London: Routledge.

Ferguson, R., 2012. *The reorder of things: the university and its pedagogies of minority difference*. Minneapolis, MN: University of Minnesota Press.

Hallward, P., 2006. *Out of this world: Deleuze and the philosophy of creation*. New York City, NY: Verso.

Lorde, A., 1993. *Zami; Sister outsider; Undersong*. New York City, NY: Quality Paperback Book Club.

Nigianni, C. and Storr, M. eds., 2009. *Deleuze and queer theory*. Edinburgh: Edinburgh University Press.

Ong, A. and Collier, S. eds., 2005. *Global assemblages: technology, politics, and ethics as anthropological problems*. Malden, MA: Blackwell.

Ross, K., 2002. *May '68 and its afterlives*. Chicago, IL: University of Chicago Press.

Ruffalo, D.V., 2009. *Post-queer politics*. Surrey: Ashgate.

Semetsky, I., 2001. The adventures of a postmodern fool; or, the semiotics of learning. In: C.W. Spinks, ed., *Trickster and ambivalence: the dance of differentiation*. Madison, WI: Atwood Publishing, pp. 57–70.

Semetsky, I., 2010. 'Silent discourse: the language of signs and "Becoming-Woman"'. *SubStance*, 39(1), pp. 87–102.

White, M., 2009. 'The master's tools: reclaiming Audre Lorde's revolutionary phrase'. *Adbusters*. Available at: https://www.adbusters.org/blogs/blackspot-blog/audre_lorde.html [Accessed 21 Mar. 2015].

Index

For Product Safety Concerns and Information please contact our EU representative GPSR@taylorandfrancis.com Taylor & Francis Verlag GmbH, Kaufingerstraße 24, 80331 München, Germany

Batch number: 10725973

Printed by Printforce, the Netherlands